O Mg! How Chemistry Came to Be

O Mg! How Chemistry Came to Be

Stephen M. Cohen

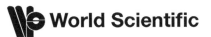
World Scientific

NEW JERSEY · LONDON · SINGAPORE · BEIJING · SHANGHAI · HONG KONG · TAIPEI · CHENNAI · TOKYO

Published by

World Scientific Publishing Co. Pte. Ltd.

5 Toh Tuck Link, Singapore 596224

USA office: 27 Warren Street, Suite 401-402, Hackensack, NJ 07601

UK office: 57 Shelton Street, Covent Garden, London WC2H 9HE

British Library Cataloguing-in-Publication Data
A catalogue record for this book is available from the British Library.

O MG! HOW CHEMISTRY CAME TO BE

ISBN 978-981-125-040-8 (hardcover)
ISBN 978-981-125-041-5 (ebook for institutions)
ISBN 978-981-125-042-2 (ebook for individuals)

For any available supplementary material, please visit
https://www.worldscientific.com/worldscibooks/10.1142/12670#t=suppl

Printed in Singapore

FOR MY FATHER, DR. PAUL S. COHEN (Z"L), WHO OUGHT TO HAVE
BEEN A CO-AUTHOR OF THIS BOOK. I KNOW YOU'D HAVE
ENJOYED CREATING IT AS MUCH AS I DID.

PHOTO COPYRIGHT © 1983 SCOTT BRENNER. USED WITH PERMISSION.

A TIP OF THE SAFETY GOGGLES TO DR. HUBERT ALYEA (1903–96),
WHOSE 1975 LECTURE ON "WHAT IS RELEVANT?" AT MT. HOLYOKE
COLLEGE SPARKED MY CHILDHOOD SELF, AND AFFECTED THIS BOOK.

AND FOR MY WIFE, WHO READ EACH PAGE AS SOON AS THE
PHOTOSHOPPED PIXELS DRIED ON THE SCREEN.

ACKNOWLEDGMENTS

MANY THANKS TO RACHEL FIELD KUTZIK, WHO JUMPSTARTED THIS PROJECT.

A VARIETY OF PEOPLE OFFERED BOTH SOLICITED AND UNSOLICITED ADVICE UPON SEEING THE MANUSCRIPT IN VARIOUS STAGES OF COMPLETION:
SASHA COHEN IOANNIDES, DR. DAVID RUBEN, SUSAN RUBEN, DR. NAOMI BASICKES, DR. REBECCA GANETZKY, CARYN ALTER, EMILY FISH, MARCY FISH, ANNA KITCES.

GIVING THE MULTITUDE OF HISTORICAL FIGURES A VOICE—A PHRASE—IN THEIR NATIVE LANGUAGES IS NO SMALL MATTER. I WANT TO THANK THE FOLLOWING PEOPLE FOR ANSWERING MY QUESTIONS ABOUT LANGUAGES:
WILLIAM BEITMANN, FRANCINE SAFIR, KARIN HOERHOLD, RAINER RICHTER, AKIVA WEINGARTEN, SASHA COHEN IOANNIDES, CAROL SCHOENLEBER, JANE VIGON, DR. RINA COHEN MULLER, PIERRE BERNY, JULLIETTE MOR-BERNY, RACHEL CHAFFIN, MICHAEL PAPAGEORGIOU, PAULA TEITELBAUM.

CONTENTS

NEVER TRUST ATOMS...

THEY MAKE UP EVERYTHING!

IGNORE HIM. START THE BOOK ALREADY!

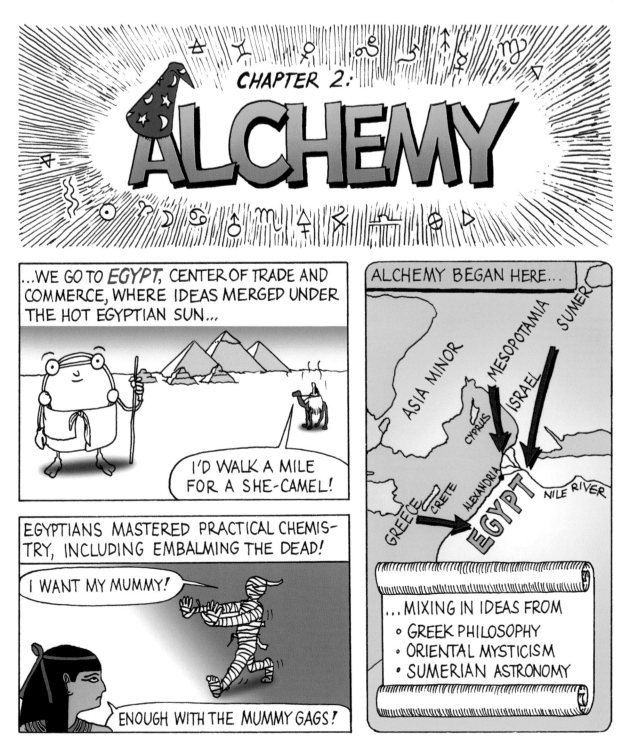

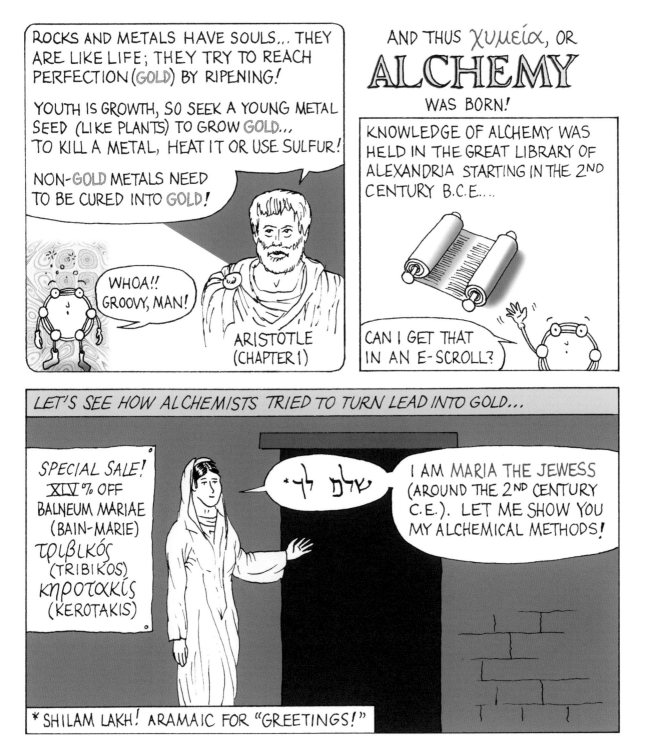

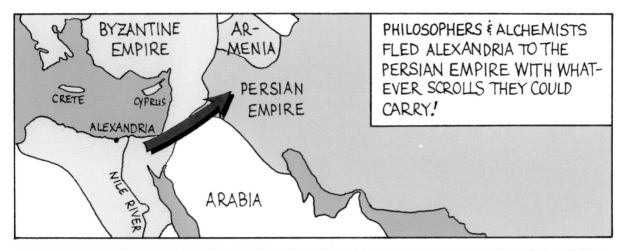

PHILOSOPHERS & ALCHEMISTS FLED ALEXANDRIA TO THE PERSIAN EMPIRE WITH WHATEVER SCROLLS THEY COULD CARRY!

IN THE BYZANTINE EMPIRE, ONE OF THE LAST USES OF GREEK ALCHEMY OCCURED WHEN CONSTANTINOPLE REPELLED ARABS IN THE 670S WITH **GREEK FIRE...**

PROBABLY A LIQUID MIXTURE OF PINE RESINS AND QUICKLIME — THE EXACT RECIPE IS LOST...

THE RISE OF ISLAM AND ARAB EMPIRES IN THE 7TH CENTURY PRESERVED WHAT REMAINED OF OLD ALCHEMY...

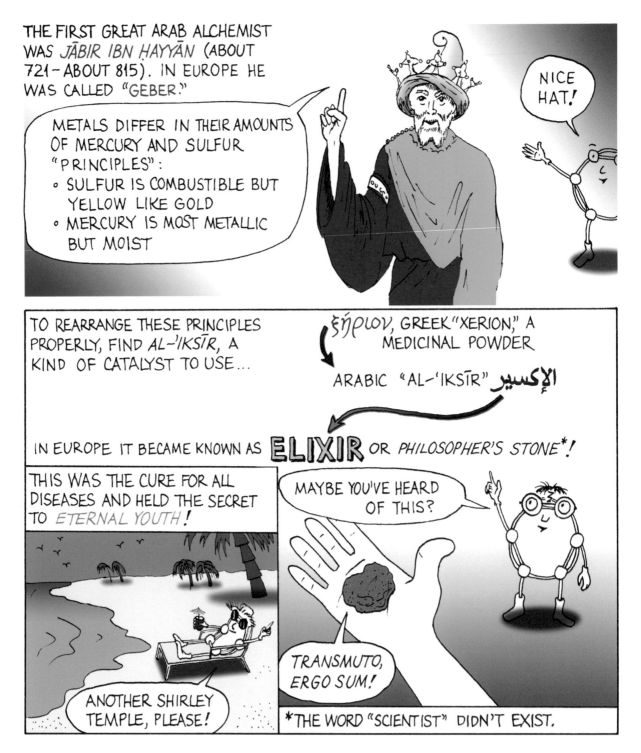

THE FIRST GREAT ARAB ALCHEMIST WAS *JĀBIR IBN ḤAYYĀN* (ABOUT 721 – ABOUT 815). IN EUROPE HE WAS CALLED "GEBER."

METALS DIFFER IN THEIR AMOUNTS OF MERCURY AND SULFUR "PRINCIPLES":
- SULFUR IS COMBUSTIBLE BUT YELLOW LIKE GOLD
- MERCURY IS MOST METALLIC BUT MOIST

NICE HAT!

TO REARRANGE THESE PRINCIPLES PROPERLY, FIND *AL-'IKSĪR*, A KIND OF CATALYST TO USE...

ξήριον, GREEK "XERION," A MEDICINAL POWDER

ARABIC "AL-'IKSĪR" الإكسير

IN EUROPE IT BECAME KNOWN AS **ELIXIR** OR *PHILOSOPHER'S STONE**!

THIS WAS THE CURE FOR ALL DISEASES AND HELD THE SECRET TO *ETERNAL YOUTH*!

MAYBE YOU'VE HEARD OF THIS?

TRANSMUTO, ERGO SUM!

ANOTHER SHIRLEY TEMPLE, PLEASE!

*THE WORD "SCIENTIST" DIDN'T EXIST.

A SECOND GREAT ARAB ALCHEMIST (AND PHYSICIAN) WAS *ABŪ BAKR MUHAMMAD IBN ZAKARIYYĀ AL-RĀZĪ* (ABOUT 825 TO ABOUT 925). TO EUROPEANS, HE WAS "RHAZES".

LET'S GO FURTHER THAN JĀBIR'S TWO PRINCIPLES (MERCURY & SULFUR).... WE'LL USE *THREE*:

MERCURY ☿ VOLATILE, METAL

SULFUR 🜍 FLAMMABLE

SALT 🜔 WHICH IS NEITHER VOLATILE NOR FLAMMABLE

SO NOW WE HAVE...

the FOUR elements

the three principles

CAN WE MAKE GOLD YET?

DESPITE— OR BECAUSE OF—THESE ALCHEMICAL IDEAS, ARAB ALCHEMISTS ADVANCED PRACTICAL CHEMISTRY BY INVENTING OR DISCOVERING:

PLASTER OF PARIS   ANTIMONY METAL

CONCENTRATED ACETIC ACID

CAUSTIC ALKALI   SAL AMMONIAC

STEEL ALLOYS

CLASSIFICATION OF SUBSTANCES BY ORIGIN

- ANIMAL
- VEGETABLE
- MINERAL
- OTHER SOURCES
  - SPIRITS
  - METALS
  - BORAXES
  - STONES
  - SALTS

DURING THE CRUSADES, THE ARABS INVADED EUROPE, IN-CREASING CON-TACT WITH IDEAS FROM THE EAST.

The Arabs are coming! The Arabs are coming!

BIG DEAL! I'M AN ARABIAN STEED!

ARAB TRADERS VISITED MANY PARTS OF EUROPE AND ASIA, DISTRIBUTING AND SHARING INTEREST-ING IDEAS AND OBJECTS...

BUT THIS "ZERO" MEANS *NOTHING*?

IT TRULY IS NOTHING, MY FRIEND!

WITH ARABIAN INFLUENCE, ALL SORTS OF INTERESTING THINGS CAME FROM

CHINA...

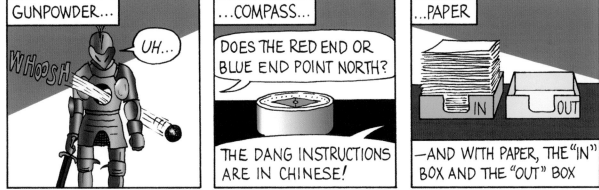

GUNPOWDER...

WHOOSH

UH...

...COMPASS...

DOES THE RED END OR BLUE END POINT NORTH?

THE DANG INSTRUCTIONS ARE IN CHINESE!

...PAPER

IN

OUT

—AND WITH PAPER, THE "IN" BOX AND THE "OUT" BOX

THE MEDIEVAL PERIOD IN EUROPE HAS A REPUTATION FOR STAGNATION AND IGNORANCE, YET PRACTICAL CHEMISTRY ADVANCED. LET'S SEE SOME EXAMPLES...

IN THE 9TH CENTURY ALCHEMISTS DISTILLED A SOLUTION OF ALCOHOL AND WATER. THEY CALLED IT A LIQUID THAT BURNS WITHOUT IGNITING OTHER THINGS:

YOUR ROASTED QUAILS, FLAMBÉED WITH COGNAC, GOOD SIR!

BUT NO FORK! WE EAT WITH OUR HANDS!

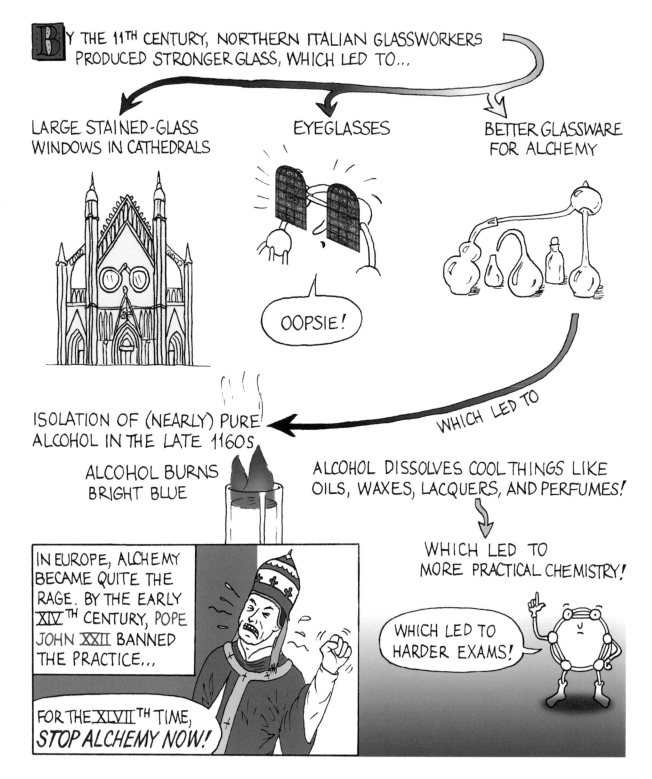

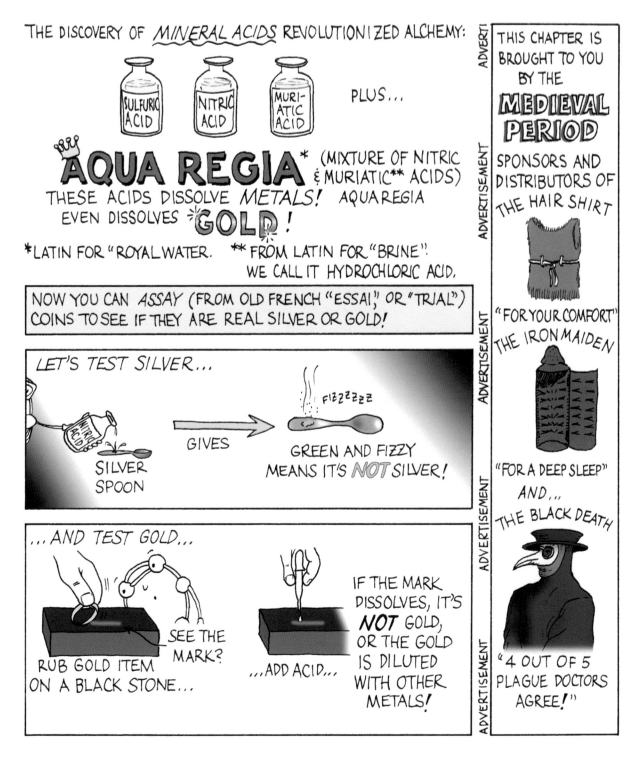

THE DISCOVERY OF *MINERAL ACIDS* REVOLUTIONIZED ALCHEMY:

SULFURIC ACID

NITRIC ACID

MURI-ATIC ACID

PLUS...

AQUA REGIA*

(MIXTURE OF NITRIC & MURIATIC** ACIDS)

THESE ACIDS DISSOLVE *METALS!* AQUA REGIA EVEN DISSOLVES *GOLD!*

*LATIN FOR "ROYAL WATER." ** FROM LATIN FOR "BRINE". WE CALL IT HYDROCHLORIC ACID.

NOW YOU CAN *ASSAY* (FROM OLD FRENCH "ESSAI," OR "TRIAL") COINS TO SEE IF THEY ARE REAL SILVER OR GOLD!

LET'S TEST SILVER...

SILVER SPOON

GIVES

FIZZZZZZZ

GREEN AND FIZZY MEANS IT'S *NOT* SILVER!

...AND TEST GOLD...

RUB GOLD ITEM ON A BLACK STONE...

SEE THE MARK?

...ADD ACID...

IF THE MARK DISSOLVES, IT'S *NOT* GOLD, OR THE GOLD IS DILUTED WITH OTHER METALS!

ANOTHER CHEMICAL TOPPLED MEDIEVAL EUROPE'S FEUDAL SOCIETY: GUNPOWDER!

...INVENTED HUNDREDS OF YEARS BEFORE IN CHINA AND FIRST MENTIONED IN EUROPE IN 1240:

From the violence of that salt called saltpeter, so horrible a sound is made by the bursting of a thing so small...that we find it exceeding the roar of a strong thunder, & a flash brighter than the most brilliant lightning!

ROGER BACON
(ENGLISH, ABOUT 1219 TO ABOUT 1292)

EUROPEANS INVENTED CANNONS IN THE 1320s...

BOOM!

THE GREAT BENZINI

AND SUDDENLY NEW CIRCUS ACTS WERE POSSIBLE...

...PLUS WALLED CITIES WERE NO LONGER PROTECTION FROM ARMIES!

MEANWHILE, CHEMISTRY HAD TO WAIT FOR ALCHEMY TO DECLINE WHILE OTHER BRANCHES OF "NATURAL PHILOSOPHY" ADVANCED.

TAPPITY TAPPITY

DURING THE RENAISSANCE, GEORG PAWER (GERMAN, 1494-1555) BECAME AN EXPERT IN METALLURGY AND MINERALOGY...

LATIN IS THE LANGUAGE OF THE COOL KIDS, SO I'LL TRANSLATE MY SURNAME PAWER (OR "BAUER," WHICH MEANS "FARMER,") INTO *AGRICOLA!*

HERE'S MY BOOK, *DE RE METALLICA* ("ON THE NATURE OF METALS")*!*

DE RE METALLICA BECAME A FOUNDING REFERENCE IN GEOLOGY!

ANOTHER INSTALLMENT OF...

STRANGE BUT TRUE!

DE RE METALLICA WAS TRANSLATED INTO ENGLISH BY LOU HENRY HOOVER, A GEOLOGIST WHO STUDIED LATIN AT STANFORD UNIVERSITY, AND HER HUSBAND, HERBERT, WHO WAS A MINING ENGINEER...

LOU HENRY HOOVER (1874-1944)

HERBERT HOOVER (1874-1964)

YEAH, *THAT* HERBERT HOOVER, THE 31ST PRESIDENT OF THE **U.S.** OF **A.**!

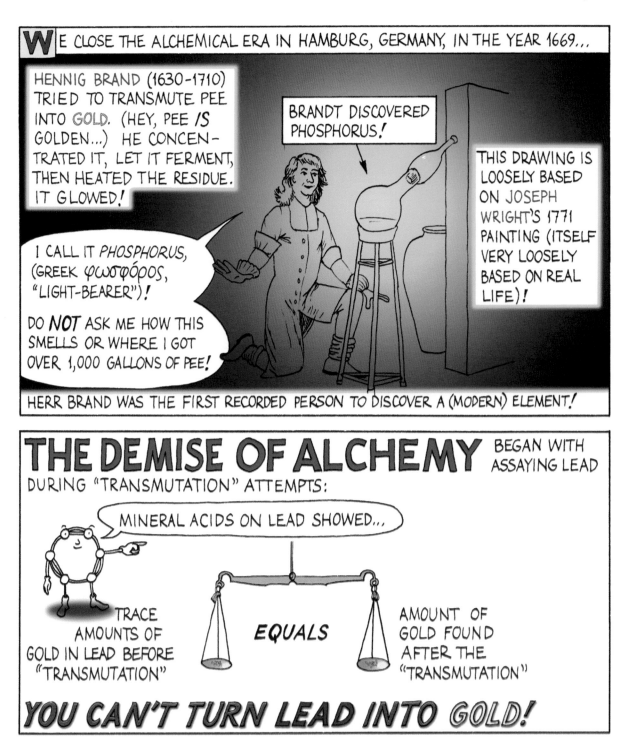

Chapter Three:
THE FIRST CHEMISTS.

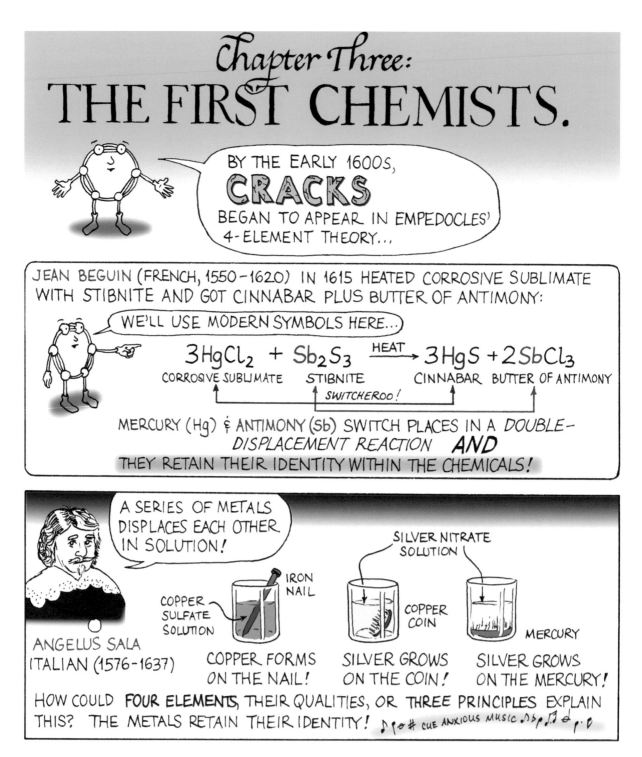

BY THE EARLY 1600s, **CRACKS** BEGAN TO APPEAR IN EMPEDOCLES' 4-ELEMENT THEORY...

JEAN BEGUIN (FRENCH, 1550–1620) IN 1615 HEATED CORROSIVE SUBLIMATE WITH STIBNITE AND GOT CINNABAR PLUS BUTTER OF ANTIMONY:

WE'LL USE MODERN SYMBOLS HERE...

$$3HgCl_2 + Sb_2S_3 \xrightarrow{\text{HEAT}} 3HgS + 2SbCl_3$$

CORROSIVE SUBLIMATE   STIBNITE   CINNABAR   BUTTER OF ANTIMONY

SWITCHEROO!

MERCURY (Hg) & ANTIMONY (Sb) SWITCH PLACES IN A *DOUBLE-DISPLACEMENT REACTION* **AND**

THEY RETAIN THEIR IDENTITY WITHIN THE CHEMICALS!

A SERIES OF METALS DISPLACES EACH OTHER IN SOLUTION!

ANGELUS SALA ITALIAN (1576–1637)

COPPER SULFATE SOLUTION

IRON NAIL

COPPER FORMS ON THE NAIL!

SILVER NITRATE SOLUTION

COPPER COIN

SILVER GROWS ON THE COIN!

MERCURY

SILVER GROWS ON THE MERCURY!

HOW COULD **FOUR ELEMENTS**, THEIR QUALITIES, OR **THREE PRINCIPLES** EXPLAIN THIS? THE METALS RETAIN THEIR IDENTITY! ♪♩♦♯ CUE ANXIOUS MUSIC ♪♭♪♫♩♪♩♪♩

G ASES BECAME THE COOL THING TO STUDY. CHECK OUT EVANGELISTA TORRICELLI'S NEW EXPERIMENT IN ITALY...

FILL A GLASS TUBE MOST OF THE WAY WITH MERCURY

36" MERCURY

CAREFULLY INVERT IT INTO A LARGE BOWL...

VACUUM!

MERCURY COLUMN FALLS TO ABOUT 30" HIGH

THIS HEIGHT VARIES A BIT

FANTASTICO!

(1608-1647)

WHY? WE LIVE SUBMERGED AT THE BOTTOM OF AN OCEAN OF AIR!* THE AIR'S WEIGHT IS PUSHING THE MERCURY UP INTO THE TUBE!

THE HEIGHT VARIES SLIGHTLY AS THE AIR PRESSURE CHANGES WITH THE WEATHER!

*HE REALLY DID WRITE THIS IN 1644 — IN ITALIAN!

OTTO VON GUERICKE (GERMANY, 1602-1686) DEMONSTRATED THE FORCE OF AIR PRESSURE DRAMATICALLY IN 1657...

METAL DOMES

GREASED JOINT

VACUUM INSIDE

OH, THE PRESSURE!

HORSES COULDN'T PULL THE DOMES APART!

AND IF YOU *STILL* DON'T BELIEVE ME, HERE ARE MORE OBSERVATIONS...

OR HOW ABOUT THIS...

FAT + ALKALI ⟶ SOAP

SOAP 🚫⟶ FAT + ALKALI

BURN WOOD TWO DIFFERENT WAYS:

OPEN GRATE ⟶ ASH + SOOT

IN A RETORT {
ASH
OIL
SPIRIT
VINEGAR
WATER
CHARCOAL
}

WHY THE DIFFERENCE?

SCIENTIFIC ASIDE: AROUND THIS TIME ANTONIE VAN LEEUWENHOEK (DUTCH, 1632-1723) DISCOVERED MICRO-ORGANISMS WITH HIS IMPROVED MICROSCOPE...

...THINGS SMALLER THAN ANYONE IMAGINED...

SO WHY NOT EVEN TINIER OBJECTS —ATOMS?

ATOMIC THEORY GAINED A FOLLOWING IN THE LATE 1600S!

BOYLE ALSO PROVIDED A PRACTICAL DEFINITION FOR ACIDS AND BASES...

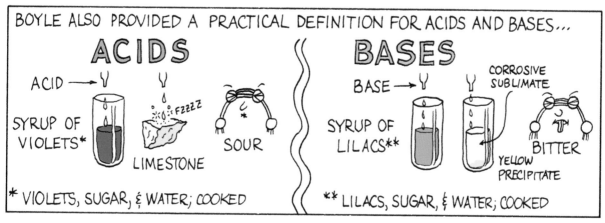

ACIDS

ACID ⟶

SYRUP OF VIOLETS*

LIMESTONE   FZZZZ

SOUR

BASES

BASE ⟶

SYRUP OF LILACS**

CORROSIVE SUBLIMATE

YELLOW PRECIPITATE

BITTER

* VIOLETS, SUGAR, & WATER; COOKED

** LILACS, SUGAR, & WATER; COOKED

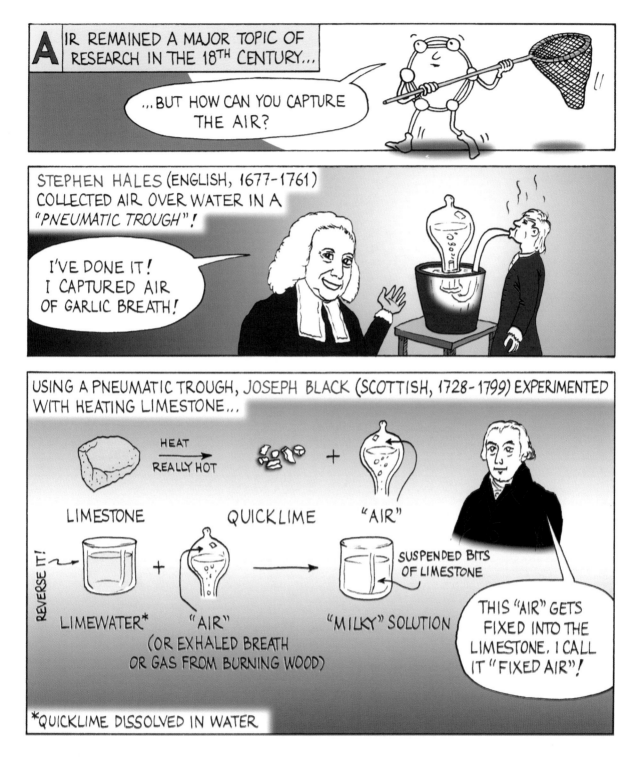

CLEARLY WHAT HAPPENED, MR. RUTHERFORD, IS THAT THE MOUSE EXHALED PHLOGISTON, THE CANDLE AND PHOSPHORUS EMITTED PHLOGISTON... UNTIL THE AIR WAS SATURATED WITH PHLOGISTON!

I AGREE, PROFESSOR BLACK! LET'S CALL THIS GAS *PHLOGISTICATED AIR!*

JOSEPH BLACK WAS SO FAMOUS FOR HIS WORK AND LECTURES THAT DR. WILLIAM WARDEN BOASTED TO NAPOLÉON BONAPARTE OF BEING BLACK'S STUDENT!

"Where," said he, "were you educated?" —I replied, "in Edinburgh."—"You have very eminent professors there I know: I remember Doctor Brown's system was in repute during my first Italian campaign. I have read of your other men of note, and I wish you would call them to my recollection by repeating their names."—I accordingly mentioned BLACK in *Chemistry*, MONRO in *Anatomy* and *Surgery*, and GREGORY in *Physic*; but, at the same time I observed, that while I particu-

Letters Written on Board H.M.S. Northumberland, 1816

LET US NOW MEET HENRY CAVENDISH (1731-1810), A RICH ENGLISH SCIENTIST WHO ALSO STUDIED AIRS...

MUST I INTRODUCE MYSELF? CAN'T I JUST SLIP YOU A NOTE AS I DO WITH MY HOUSEKEEPER?

CAVENDISH RESEARCHED THE AIR EMITTED WHEN ACIDS REACT WITH METALS...

"AIR"

ACID

IRON

THIS NEW AIR BURNED EASILY, AND IT WEIGHED ONLY 1/14 AS MUCH AS ORDINARY AIR! HE CALLED IT *INFLAMMABLE AIR!*

WAS IT PURE PHLOGISTON?

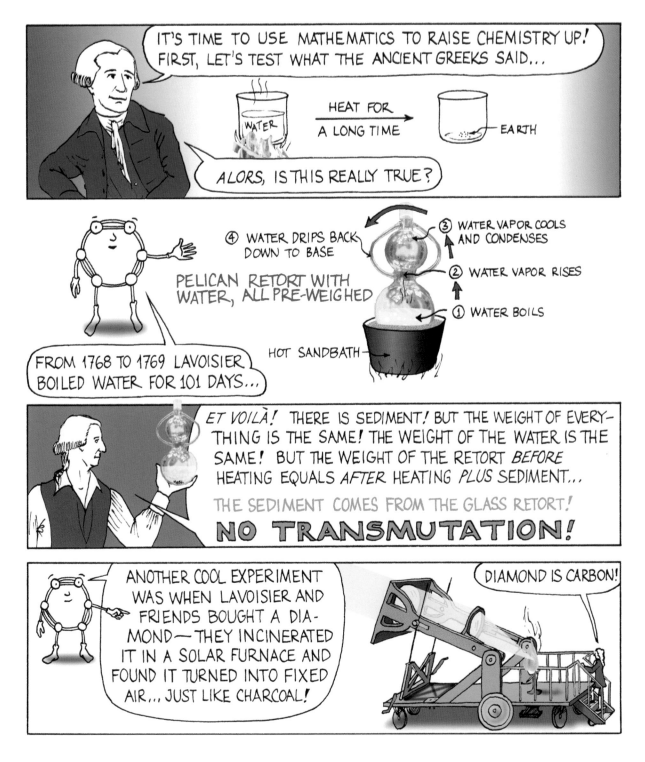

LAVOISIER STUDIED CALXES OF METALS:

AIR PLUS TIN OR LEAD IN A SEALED JAR, ALL WEIGHED

HEAT

AIR PLUS CALX PLUS TIN OR LEAD IN A SEALED JAR, ALL WEIGHED

WEIGHT IS THE SAME

METAL + CALX GAINED WEIGHT AIR LOST WEIGHT

ZUT ALORS! THE METAL COMBINES WITH SOMETHING IN THE AIR... *PHLOGISTON IS NOT INVOLVED!*

THE REVERSE REACTION *DOESN'T* INVOLVE PHLOGISTON...

ORE (METAL + GAS) + CHARCOAL → HEAT → METAL + FIXED AIR

BASED ON HIS EXPERIMENTS, LAVOISIER FIRST STATED

THE LAW OF CONSERVATION OF MASS:

MATTER IS NEITHER CREATED NOR DESTROYED*

DON'T VIOLATE IT!

ALMOST — IN THE 20TH CENTURY THIS WAS FOUND TO BE NOT QUITE TRUE!

BUT... NOT *ALL* THE AIR COMBINED WITH METAL MADE CALX — ONLY 1/5 DID...

OUR PAL, JOSEPH PRIESTLEY, VISITED LAVOISIER IN 1774...

... SO I FOUND DEPHLOGISTI- CATED AIR EMITTED FROM THE CALX OF MERCURY!

SON FRAN- ÇAIS EST AFFREUX!

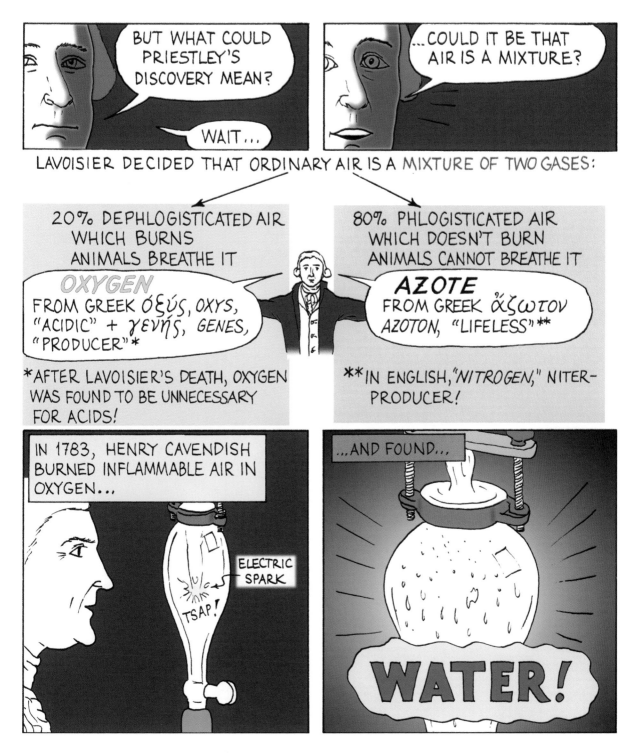

T O PROMOTE HIS NEW CHEMISTRY, LAVOISIER DID THREE THINGS:

WITH HIS FRIENDS DEMORVEAU, BERTHOLLET, AND FOURCROY, HE RATIONALIZED CHEMICAL NOMENCLATURE...

NO MORE "FLOWERS OF ZINC"! NOW IT'S "ZINC OXIDE"!

IF YOU USE MY WORDS, YOU AGREE WITH MY SYSTEM!

HE WROTE THE FIRST MODERN CHEMISTRY TEXTBOOK, *ELEMENTS OF CHEMISTRY, IN A NEW SYSTEMATIC ORDER...*

THAT'S THE ENGLISH TITLE!

HIS WIFE, MARIE-ANNE, STAGED A RITUAL BURNING OF GEORG STAHL'S BOOKS, WHILE DRESSED AS A PRIESTESS!

AU REVOIR, M. STAHL!

(FOR GEORG STAHL SEE CHAPTER 3)

FINALLY, LAVOISIER AND HIS ANTI-PHLOGISTIC FRIENDS FOUND THAT THE EXISTING FRENCH CHEMISTRY JOURNALS WERE ALL PRO-PHLOGISTON...

...SO WE FOUNDED OUR OWN PRO-OXYGEN JOURNAL...

...WHICH STILL EXISTS TODAY!

WHAT HAPPENED TO THEM?

PRIESTLEY WAS A RADICAL WHO SYMPATHIZED WITH FOREIGNERS AND THEIR REVOLUTIONS, *AND* HE WAS UNITARIAN. RIOTERS ATTACKED HIS HOME IN 1791, SO HE MOVED TO LONDON. RIOTERS ATTACKED AGAIN, SO HE FLED TO AMERICA.

HE NEVER ACCEPTED LAVOISIER'S CHEMISTRY!

YOU CAN VISIT PRIESTLEY'S HOME IN NORTHUMBERLAND, PENNSYLVANIA!

LAVOISIER COLLECTED TAXES FOR THE KING AND HAD CONTACTS WITH FOREIGNERS, SO HE WAS SUSPICIOUS TO THE LEADERS OF THE NEW FRENCH REPUBLIC. HE WAS ARRESTED IN 1793, CONVICTED, AND GUILLOTINED IN 1794!

Le laboratoire de Lavoisier

YOU CAN SEE LAVOISIER'S EQUIPMENT AT THE MUSÉE DES ARTS ET MÉTIERS, IN PARIS.

WITHIN A DECADE, LAVOISIER'S NEW CHEMISTRY WAS WIDELY ACCEPTED...

PUSH

LAVOISIER

OLD CHEMISTRY

...EXCEPT FOR SOME OLDER FOLKS LIKE PRIESTLEY AND CAVENDISH...

CRASH!

HARRUMPH! PHLOGISTON IS REAL!

ELIZABETH FULHAME'S REMARKABLE BOOK FROM 1794, *AN ESSAY ON COMBUSTION, WITH A VIEW TO A NEW ART OF DYING AND PAINTING*, SHOWS HOW RAPID THE TRANSFORMATION WAS...

PREFACE.   ix

which includes every thing the experiments can extend to.

As to the style, I have endeavoured to relate the experiments in a plain and simple manner, aiming more at perspicuity, than

I have adopted the French Nomenclature, as the terms of it are so framed, as to prevent circumlocution, assist the memory, by pointing out the combination, and state of the elements existing in each compound, as far as they are known ; advantages to be found in no other Nomenclature.

However, the English reader must regret, that the French chymists have not preferred the terms *air*, and *ammonia*, to the less harmonious sounds, *gas*, and *ammoniac*. I took the liberty of writing the latter *ammonia*.

order, in which they were made, sensible

FULHAME DESCRIBED HER EXPERIMENTS ON LIGHT-INDUCED REDUCTION OF GOLD AND SILVER, PHOTO-IMAGING ON CLOTH, AND EVEN *CATALYSIS*— WHEN A SUBSTANCE TAKES PART IN A REACTION BUT *IS NOT* USED UP !

DON'T CONFUSE THE "LONG S" WITH AN "f" !

FULHAME ALSO HAD SOME THINGS TO SAY ABOUT SEXISM IN THE 1790s...

It may appear presuming to *some*, that I should engage in pursuits of this nature, but averse from indolence, and having much leisure, my mind led me to this mode of amusement, which I found entertaining, and will, I hope, be thought inoffensive by the liberal, and the learned. But censure is perhaps inevitable ; for some are so ignorant, that they grow sullen and silent, and are chilled with horror at the sight of any thing, that bears the semblance of learning, in whatever shape it may appear ; and should the *spectre* appear in the shape of *woman*, the pangs, which they suffer, are truly dismal.

Chapter Five.
ATOMS (AGAIN)

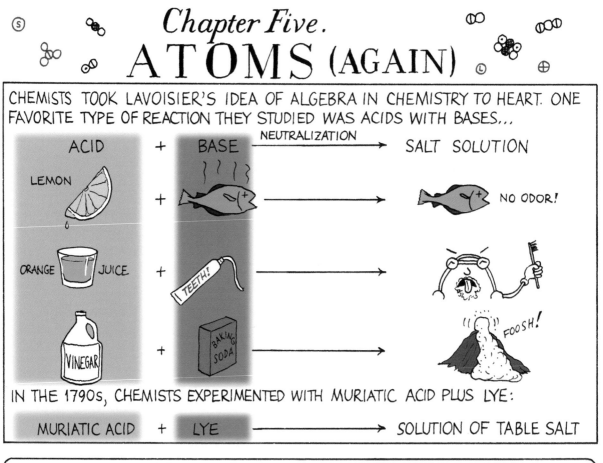

CHEMISTS TOOK LAVOISIER'S IDEA OF ALGEBRA IN CHEMISTRY TO HEART. ONE FAVORITE TYPE OF REACTION THEY STUDIED WAS ACIDS WITH BASES...

ACID + BASE → (NEUTRALIZATION) → SALT SOLUTION

LEMON + (fish) → (fish) NO ODOR!

ORANGE JUICE + TEETH! →

VINEGAR + BAKING SODA → FOOSH!

IN THE 1790s, CHEMISTS EXPERIMENTED WITH MURIATIC ACID PLUS LYE:

MURIATIC ACID + LYE → SOLUTION OF TABLE SALT

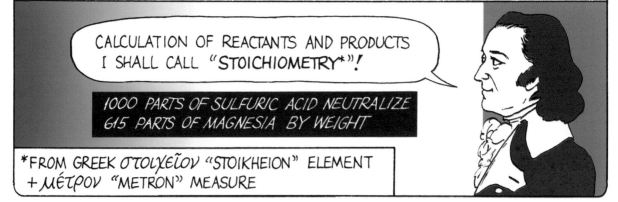

JEREMIAS BENJAMIN RICHTER (GERMAN, 1762–1807) DISCOVERED THAT EXACT AMOUNTS OF ACIDS NEUTRALIZED EXACT AMOUNTS OF BASES... HE CALLED THIS "EQUIVALENT WEIGHTS."

CALCULATION OF REACTANTS AND PRODUCTS I SHALL CALL "STOICHIOMETRY*"!

1000 PARTS OF SULFURIC ACID NEUTRALIZE 615 PARTS OF MAGNESIA BY WEIGHT

*FROM GREEK στοιχεῖον "STOIKHEION" ELEMENT + μέτρον "METRON" MEASURE

D ALTON FLESHED OUT HIS IDEAS IN HIS BOOK *A NEW SYSTEM OF CHEMICAL PHILOSOPHY* IN 1808:

1. ATOMS ARE SOLID, INDIVISIBLE, AND INCOMPRESSIBLE, SURROUNDED BY AN ATMOSPHERE OF HEAT

2. ATOMS ARE INDESTRUCTIBLE — *NO TRANSMUTATION!*

3. EACH ELEMENT HAS ITS OWN KIND OF ATOM

4. EACH ELEMENT'S ATOMS HAVE A MEASURABLE, DISTINCT WEIGHT.

R.I.P. ALCHEMY 2500 BCE TO 1808 CE

FINALLY!

DALTON'S LOGIC WENT LIKE THIS... (AND WE'LL USE HIS SYMBOLS)

WE GUESS THE SIMPLEST ATOMIC PROPORTIONS* FOR WATER, A COMPOUND:

   1 PART BY WEIGHT HYDROGEN
        +
   7 PARTS BY WEIGHT OXYGEN   → WATER

...AND FOR ANOTHER COMPOUND, AMMONIA:

   1 PART BY WEIGHT HYDROGEN
        +
   5 PARTS BY WEIGHT NITROGEN   → AMMONIA

LET'S SET HYDROGEN, THE LIGHTEST ELEMENT, = 1

                        NITROGEN = 5

                        OXYGEN = 7

*NOT ALWAYS TRUE, BUT LET'S WORK WITH HIM HERE...

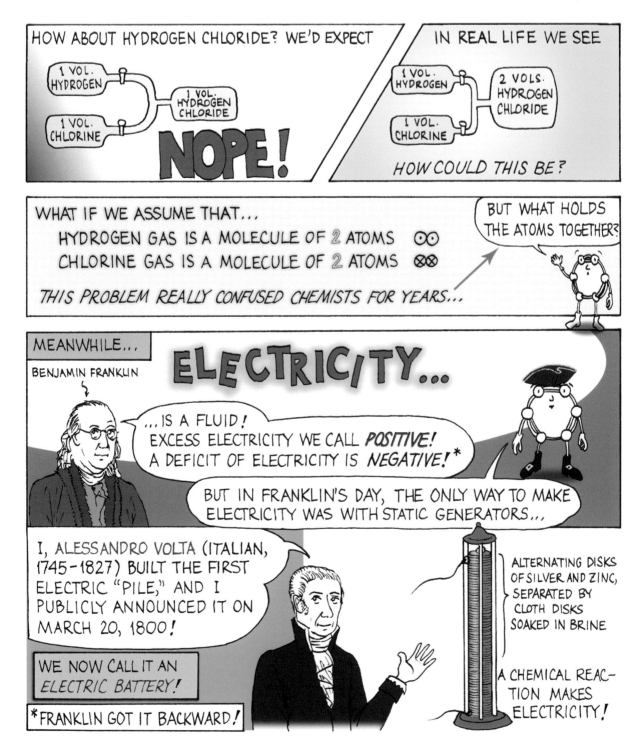

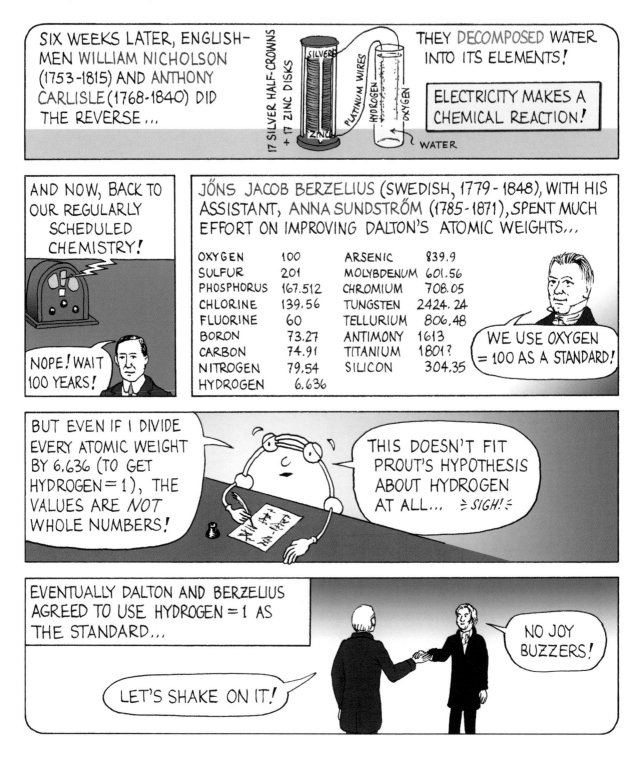

NOW THAT CHEMISTS WERE GETTING MORE COMFORTABLE WITH ATOMS, HOW DID CHEMISTS SYMBOLIZE THEM?

DALTON USED CIRCLES:

PRINTERS FREAKED OUT WHEN THEY SAW THESE SYMBOLS...

BERZELIUS REMOVED THE CIRCLES...HE USED 1 OR 2 LETTERS FROM LATIN NAMES:

GOLD *AURUM* Au   IRON *FERRUM* Fe
SILVER *ARGENTUM* Ag   TIN *STANNUM* Sn
SODIUM *NATRIUM* Na   COPPER *CUPRUM* Cu
MERCURY *HYDRARGYRUM* Hg
LEAD *PLUMBUM* Pb

GUESS WHOSE SYMBOLS THE PRINTING HOUSES PREFERRED?

COMPARE THE TWO WAYS OF DEPICTING MOLECULES:

CO CARBON MONOXIDE
CO_2 CARBON DIOXIDE
H_2O WATER
H_2SO_4 SULFURIC ACID

WITHIN A FEW DECADES, *MY* SYMBOLS WERE ADOPTED BY CHEMISTS AROUND THE WORLD!

CHEMICAL EQUATIONS REPRESENT REACTIONS:

$$C + O_2 \longrightarrow CO_2$$
$$H_2 + Cl_2 \longrightarrow 2HCl \quad \text{(2 MOLECULES OF HCl ARE FORMED)}$$
$$2H_2 + O_2 \longrightarrow 2H_2O$$

THE SAME NUMBER AND TYPE OF ATOM MUST BE ON BOTH SIDES!

A "BALANCED EQUATION"!

AND THE NUMBER OF ELEMENTS GREW...

Mo W Y Zr Be Sc U Nb

IN NATURE, MANY ELEMENTS ARE COMBINED WITH OXYGEN, AS AN *OXIDE* (OR "CALX" IN PRE-LAVOISIER TERMINOLOGY):

IRON ORE (IRON OXIDES)   CASSITERITE (TIN OXIDE)

PYROLUSITE (MANGANESE OXIDE)   ZINCITE (ZINC OXIDE)

CUPRITE (COPPER OXIDE)   CHROMITE (CHROMIUM + IRON OXIDE)

TO ISOLATE THE ELEMENT, YOU HAVE TO REMOVE THE OXYGEN...
FIND ANOTHER ELEMENT WITH A STRONGER AFFINITY* FOR OXYGEN!

IRON OXIDE   +   CHARCOAL   →   GASES CO, CO_2 +   IRON
CARBON OXIDES

HERE WE USE CARBON TO GRAB OXYGEN FROM THE IRON!

*NO ONE KNEW WHAT CAUSED THIS "AFFINITY"!

QUICKLIME SEEMS TO BE AN OXIDE...
... BUT AN OXIDE OF *WHAT*??

NO ONE KNEW...
...BECAUSE NOTHING WOULD PULL THE OXYGEN OFF OF QUICKLIME!

ENTER HUMPHRY DAVY (ENGLISH, 1778 - 1829)...

LET'S TRY THIS COOL, NEW ELECTRIC BATTERY!

SO *THAT'S* WHY WE MENTIONED VOLTA'S PILE!

DAVY STACKED 250 METAL DISKS TO BUILD A BATTERY. HE TRIED THIS BATTERY ON SOLUTIONS OF OXIDES... BUT ONLY DECOMPOSED THE WATER!

WHAT IF I SKIPPED THE WATER AND *MELTED* THE OXIDES?

IT WORKED!

DAVY MELTED CAUSTIC POTASH IN 1807...

CAUSTIC POTASH FLAKES → POTASSIUM METAL!

THEN HE MELTED LYE...

LYE PELLETS → SODIUM METAL!

DAVY TURNED VOLTA'S BATTERY SIDEWAYS

AND WITHIN A YEAR, HE DISCOVERED...

BORON, MAGNESIUM, STRONTIUM, BARIUM, AND CALCIUM—THAT UNKNOWN METAL IN QUICKLIME!

DAVY HELD THE RECORD FOR MOST ELEMENTS DISCOVERED FOR 150 YEARS!

DAVY EVEN SHOWED THAT "OXYMURIATIC ACID" WAS REALLY HYDROCHLORIC ACID, WITH *NO* OXYGEN. DAVY DISPROVED LAVOISIER'S ASSERTION THAT *ALL* ACIDS CONTAIN OXYGEN!

TAKE THAT, *MON AMI!*

HUMPHRY DAVY WAS A POPULAR LECTURER DEMONSTRATING CHEMISTRY AT THE ROYAL INSTITUTION...

DAVY

ONE AUDIENCE MEMBER WAS JANE MARCET (1769–1858), A WRITER...

HERE'S HOW...

PART ①
$$FeS_2 + C \longrightarrow CS_2 + Fe$$
$$CS_2 + 2Cl_2 \longrightarrow CCl_4 + 2S$$
$$2CCl_4 \longrightarrow C_2Cl_4 + 2Cl_2$$

PART ②
$$2H_2 + O_2 \longrightarrow 2H_2O$$

PART ③
$$C_2Cl_4 + 2H_2O + Cl_2 \longrightarrow CCl_3CO_2H + 3HCl$$
$$CCl_3CO_2H + 3H_2 \longrightarrow \mathbf{CH_3CO_2H} + 3HCl$$

NO ORGANIC STARTING MATERIALS!

AND THEN PIERRE MARCELLIN BERTHELOT (FRENCH, 1827–1907) SYNTHESIZED A BUNCH OF ORGANIC COMPOUNDS...

ETHANOL, METHANE, FORMIC ACID, ACETYLENE...

AND EVEN *BENZENE!*

WHOA! THAT'S MY ANCESTOR!

BUT UNDERSTANDING LARGE MOLECULES WAS STILL A PROBLEM:

LIKE STARCH, FAT, AND PROTEIN!

STILL, CHEMISTS MADE SOME PROGRESS...

GOTTLIEB KIRCHHOFF HEATED STARCH WITH SULFURIC ACID AND GOT GLUCOSE (A SUGAR)!

♪ I HEAT UP, I CAN'T COOL DOWN... ♪

HENRI BRACONNOT HEATED GELATIN WITH ACID AND FOUND THE AMINO ACID GLYCINE, THEN HEATED MUSCLE AND GOT ANOTHER AMINO ACID, LEUCINE!

MICHEL CHEVREUL SHOWED THAT FATS AND OILS WERE COMBINATIONS OF ORGANIC ACIDS PLUS GLYCEROL...

BERTHELOT HEATED GLYCEROL WITH STEARIC ACID (A COMMON FATTY ACID), AND GOT TRISTEARIN!

BY 1854 HE REACTED OTHER ACIDS WITH GLYCEROL AND GOT TRIACETIN, A COMPOUND NOT FOUND IN NATURE!

$C_9H_{14}O_6$!

HE MADE THE FIRST *ARTIFICIAL* ORGANIC COMPOUND!

TAKE *THAT*, BERZELIUS!

AUGUST KEKULÉ'S 1859 *LEHRBUCH DER ORGANISCHEN CHEMIE...*

"SO WE DEFINE ORGANIC CHEMISTRY AS THE CHEMISTRY OF CARBON COMPOUNDS."

THIS OLD TOME DESCRIBES MY FAMILY!

zusammen zu fassen.
Wir definiren also die organische Chemie als die Chemie der Kohlenstoffverbindungen. Wir sehen dabei keinen Gegensatz zwischen unorganischen und organischen Verbindungen. Das was

IN THE EARLY 19TH CENTURY, CHEMISTS ANALYZED COMPOUNDS ONLY FOR THEIR *EMPIRICAL FORMULA:* THE PROPORTIONS OF EACH ELEMENT IN A COMPOUND!

BUT IN THE MID-1820s...

JUSTUS VON LIEBIG (GERMAN, 1803-1873)

WÖHLER (AGAIN)

SILVER FULMINATE!

SILVER CYANATE!

ITS FORMULA IS AgCNO! *SNAP!*

BERZELIUS FOUND THAT TWO DIFFERENT ORGANIC ACIDS, TARTARIC ACID AND RACEMIC ACID, WERE BOTH $C_4H_6O_6$!

I SHALL CALL THEM *ISOMERS* (FROM GREEK ισομερής, "EQUAL PART")!

NOW THE QUESTION IS, HOW ARE ATOMS ARRANGED IN A MOLECULE?

?

THE FIRST BREAK IN THE CASE CAME IN THE 1810s...

NOPE! NOT FOR ANOTHER 75 YEARS...

GAY-LUSSAC AND HIS COLLEAGUE LOUIS-JACQUES THÉNARD (1777-1857) STUDIED CYANIC ACID, HCN! THEY FOUND THAT CN (CYANIDE) STAYED AS A UNIT DURING REACTIONS:

$$HCN + Cl_2 \longrightarrow HCl + ClCN$$

THEY ALSO FOUND THAT THESE SALTS WERE SIMILAR:

NaCN ☠ BUT *DEATHLY POISONOUS!*)
NaI
NaCl   (TABLE SALT)

IN ORGANIC CHEMISTRY, THEY FOUND THAT SUGAR, STARCH, AND WOOD ALL HAD PROPORTIONS OF HYDROGEN AND OXYGEN AS WATER! THESE MATERIALS SEEMED TO BE **HYDRATES** OF CARBON, OR...

ＣＡＲＢＯＨＹＤＲＡＴＥＳ $C(H_2O)$

BY 1832 LIEBIG AND WÖHLER WERE FRIENDS... THEY DISCOVERED BENZOYL (C_7H_5O), WHICH ALSO ACTED AS A UNIT...

$$BENZOYL + Cl \longrightarrow C_7H_5OCl$$

BENZOYL CHLORIDE (LIKE HCl!)

CHEMISTS CALLED THESE MULTI-ATOM UNITS *RADICALS* (LATIN *RADIX*, "ROOT").

MAYBE *ALL* ORGANIC COMPOUNDS ARE BUILT FROM RADICALS?

JA, *MEIN FREUND!*

TOTALLY RADICAL, DUDE!

FRENCHMEN JEAN-BAPTISTE DUMAS (1800-1884) AND FÉLIX-POLYDORE BOULLAY (1806-1835) PROPOSED THAT ETHYLENE (C_2H_4) IS A RADICAL, CALLED *ETHERIN*...

SO, $C_2H_4 + H_2O$ IS C_2H_6O (ETHANOL)
$C_2H_4 + HCl$ IS SALT ETHER
$2C_2H_4 + H_2O$ IS ETHER
EVEN AMMONIA, NH_3 ACTS THIS WAY...
SO THAT $NH_3 + HCl$ IS SALT AMMONIA
(WE CALL IT AMMONIUM CHLORIDE)

ACCORDING TO CHARLES ADOLPHE WURTZ (FRENCH, 1817-84)...

AMINES ARE AN "AMMONIA" TYPE!

$$H-\!\!\!\begin{array}{c}H\\|\\N\\|\\H\end{array} \qquad C_2H_5-\!\!\!\begin{array}{c}H\\|\\N\\|\\H\end{array} \qquad \begin{array}{c}C_2H_5\\ \diagdown\\ C_2H_5-N\\ |\\ H\end{array} \qquad \begin{array}{c}C_2H_5\\ \diagdown\\ C_2H_5-N\\ |\\ C_2H_5\end{array} \qquad \begin{array}{c}C_2H_5\\ \diagdown\\ C_2H_5-N\cdot I\\ \diagup |\\ C_2H_5 \quad C_2H_5\end{array}$$

AMMONIA   ETHYLAMINE   DIETHYLAMINE   TRIETHYLAMINE   TETRAETHYL-AMMONIUM IODIDE

ALL THIS TALK OF "TYPES" AND "SUBSTITUTION" OF ATOMS GOT WÖHLER TO WRITE A PRANK LETTER, WHICH LIEBIG PUBLISHED IN 1840...

① PUT CHLORINE GAS INTO A SOLUTION OF MANGANESE ACETATE ($MnO + C_4H_6O_3$) IN SUNLIGHT, EVENTUALLY YOU GET $Mn_2Cl + C_4Cl_6O_3$! HA HA!

② ADD MORE CHLORINE GAS TO THE SALT, SO THAT *ALL* H's AND MnO ARE REPLACED BY Cl's, GIVING $Cl_2Cl_2 + C_4Cl_6O_3$! HA HA!

③ THEN DISSOLVE IT IN WATER; CO_2 GAS BUBBLES OFF, AND YOU GET $Cl_2Cl_2 + Cl_8Cl_6Cl_6$! HEH!

SIGNED   *S.C.H. WINDLER*
BAHA HAHAHAHA!

HERE IS HOW CHEMISTS WROTE WATER-TYPE MOLECULES:

$$\left.\begin{array}{c}H\\H\end{array}\right\}O \qquad \left.\begin{array}{c}C_2H_5\\H\end{array}\right\}O \qquad \left.\begin{array}{c}C_2H_5\\C_2H_5\end{array}\right\}O \qquad \left.\begin{array}{c}K\\K\end{array}\right\}O \qquad \left.\begin{array}{c}K\\H\end{array}\right\}O \leftarrow \text{THE LINK IS OXYGEN}$$

WATER   ETHANOL   DIETHYL ETHER   POTASSIUM OXIDE   POTASSIUM HYDROXIDE

CHARLES FRÉDÉRIC GERHARDT (FRENCH, 1816-56) ADDED TO THE THEORY OF TYPES...

THERE ARE 4 TYPES...

① LEAST COMPLEX
H_2 HYDROGEN

② MORE COMPLEX
HCl   HYDROGEN CHLORIDE

③ EVEN MORE COMPLEX
H_2O WATER

④ MOST COMPLEX
NH_3 AMMONIA

(BUT WE STILL *DON'T KNOW ATOMS' ARRANGEMENT...*)

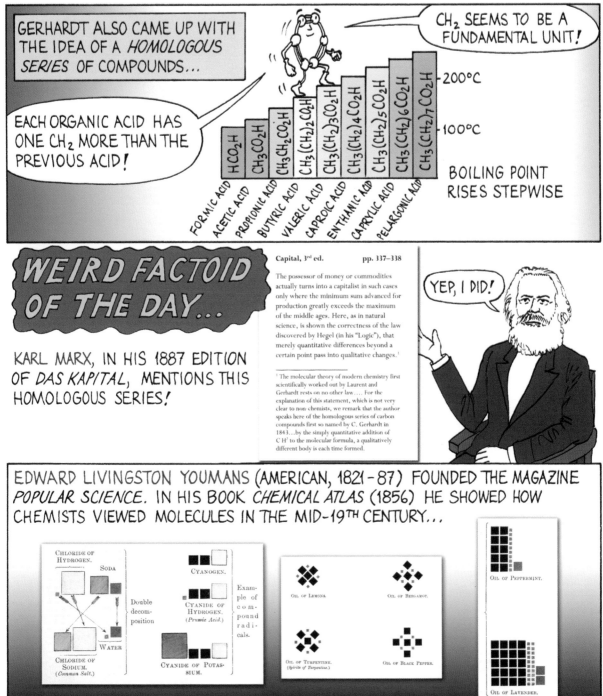

EDWARD FRANKLAND (ENGLISH, 1825-99) STUDIED COMPOUNDS FORMED WHEN ZINC METAL REACTS WITH ETHYL IODIDE (C_2H_5I)...

THIS TUBE CONTAINS DIETHYL ZINC, $(C_2H_5)_2Zn$!

FRANKLAND BEGAN THE STUDY OF *ORGANOMETALLIC* COMPOUNDS— HAVING AN ORGANIC *AND* A METAL PART!

PERHAPS THE MOST FAMOUS ONE IS VITAMIN B_{12}!

BUT CHECK *THIS* OUT: ZINC *ALWAYS* COMBINES WITH 2 EQUIVALENTS OF OTHER SUBSTANCES!

FRANKLAND NOTICED THAT MANY ELEMENTS HAVE A FIXED COMBINING POWER, WHICH CAME TO BE CALLED **VALENCE** (FROM LATIN *VALENTIA*, "POWER, CAPACITY")

VALENCE	ELEMENT
1	H, Na, Cl, Ag, Br, K
2	O, Ca, S, Mg, Ba, Zn
3	Al, Au, Sb, As
2 or 3	Fe
3 or 5	N, P

IN 1857, AUGUST KEKULÉ ADDED CARBON WITH VALENCE 4 TO THE LIST...

KNOWING THE VARIOUS VALENCES OF DIFFERENT ELEMENTS LED TO...

UM,... WAIT... I'M GETTING WORD THAT IT'S TIME FOR A SPECIAL CHAPTER ON...

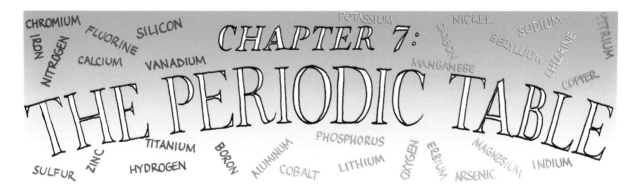

CHAPTER 7: THE PERIODIC TABLE

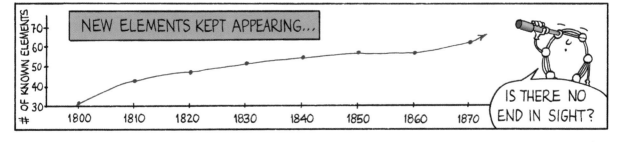

NEW ELEMENTS KEPT APPEARING...

OF KNOWN ELEMENTS

70 60 50 40 30

1800 1810 1820 1830 1840 1850 1860 1870

IS THERE NO END IN SIGHT?

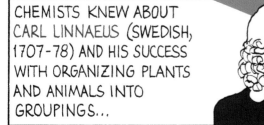

CHEMISTS KNEW ABOUT CARL LINNAEUS (SWEDISH, 1707-78) AND HIS SUCCESS WITH ORGANIZING PLANTS AND ANIMALS INTO GROUPINGS...

YOU CAN USE MY SYSTEM— AS LONG AS YOU NAME AN ELEMENT AFTER ME!

...LIKE *THAT* WOULD HAPPEN...

IN 1829, JOHANN WOLFGANG DÖBEREINER (GERMAN, 1780-1849) NOTICED TRIADS OF CHEMICALLY SIMILAR ELEMENTS... THE MIDDLE ELEMENT OF EACH TRIAD HAD AN ATOMIC WEIGHT NEAR THE AVERAGE OF THE OTHER TWO!

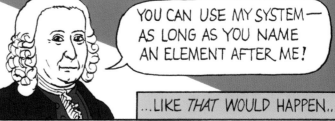

SALT-FORMERS

CHLORINE	35.5	
BROMINE	78.3	(AVERAGE = 81)
IODINE	126.5	

ALKALINE-EARTH FORMERS

CALCIUM	40.1	
STRONTIUM	87.6	(AVERAGE = 88.7)
BARIUM	137.3	

ALKALI-FORMERS

LITHIUM	6.9	
SODIUM	23.0	(AVERAGE = 23.0)
POTASSIUM	39.1	

ACID-FORMERS

SULFUR	32.2	
SELENIUM	79.3	(AVERAGE = 80.7)
TELLURIUM	129.2	

BUT NO ONE FOUND MORE TRIADS!

IN 1857, WILLIAM ODLING (ENGLISH, 1829-1921) SORTED THE ELEMENTS INTO 13 GROUPS BASED ON A VARIETY OF PHYSICAL AND CHEMICAL PROPERTIES:

GROUP	ELEMENTS
1	F, Cl, Br, I
2	O, S, Se, Te
3	N, P, As, Sb, Bi
4	B, Si, Ti, Sn
5	Li, Na, K
6	Ca, Sr, Ba
7	Mg, Zn, Cd
8	Be*, Y, Th
9	Al, Zr, Ce, U
10	Cr, Mn, Co, Fe, Ni, Cu
11	Mo, V, W, Ta
12	Hg, Pb, Ag
13	Pd, Pt, Au

CHECK OUT MY CHART!

BUT IT STILL BEATS THE HECK OUTTA ME AS TO HOW MANY MORE ELEMENTS THERE ARE!

*ODLING USED Gl (GLUCINIUM), AN OLD NAME!

CHEMISTS WERE STILL CONFUSED ABOUT THE MEANINGS OF EQUIVALENT WEIGHT, MOLECULAR WEIGHT, AND ATOMIC WEIGHT...

FOR EXAMPLE, OXYGEN:
- EQUIVALENT WEIGHT = 8 (COMBINES WITH 1/8 ITS WEIGHT IN HYDROGEN)
- ATOMIC WEIGHT = 16 (16 TIMES THE WEIGHT OF A HYDROGEN ATOM)
- MOLECULAR WEIGHT IS 32 (O_2 = 32)

KEKULÉ HAD ENOUGH OF THIS...

OKAY, LET'S HAVE A MEETING OF CHEMISTS IN KARLSRUHE, GERMANY!

WE'LL HASH ALL THIS OUT!

THE FIRST INTERNATIONAL CHEMICAL CONGRESS, WITH 127 DELEGATES, MET ON SEPTEMBER 3, 1860...

T-SHIRT THANKS TO ARTHUR GREENBERG, A CHEMICAL HISTORY TOUR, 2000.

ON THE LAST DAY OF THE CONFERENCE, ONE OF THE DELEGATES, ANGELO PAVESI, HANDED OUT BROCHURES BY STANISLAO CANNIZZARO (ITALIAN, 1826-1910). IN THEM, CANNIZZARO PROMOTED AN IDEA BY AMEDEO AVOGADRO (ITALIAN, 1776-1856) 50 YEARS EARLIER...

EQUAL NUMBERS OF GAS PARTICLES HAVE THE SAME VOLUME! (AT THE SAME TEMPERATURE)

AVOGADRO, NOT CANNIZZARO

MOLECULES WITH 1, 2, 3, OR 4 ATOMS...

MERCURY VAPOR

OXYGEN GAS

WATER VAPOR

AMMONIA GAS

THE DELEGATES READ THE PAMPHLET...

I WAS ASTONISHED AT ITS CLARITY... IT WAS AS IF SCALES FELL FROM MY EYES, DOUBTS VANISHED, AND A FEELING OF CALM CERTAINTY CAME OVER ME!

HALLO MEIN NAME IST LOTHAR MEYER

DELEGATES MEET HERE

NOW THAT CHEMISTS AGREED ON ATOMIC AND MOLECULAR WEIGHTS, ALEXANDRE-ÉMILE-BEGUYER DE CHANCOURTOIS (FRENCH, 1820-86) USED THEM TO CLASSIFY ELEMENTS...

ET VOILÀ! MY VIS TELLURIQUE*!

BUT THE JOURNAL NEVER PUBLISHED AN IMAGE OF THIS—SO HE WAS IGNORED!

*TELLURIC SCREW

WHILE WRITING, MENDELEEV WOULD LAY OUT CARDS WITH THE ELEMENTS ON THEM IN ORDER OF THEIR PROPERTIES...

THEY'RE IN ORDER OF ATOMIC WEIGHT!

HE ANNOUNCED THE RESULT IN MARCH 1869 AS THE "**PERIODIC SYSTEM**"...

HE LEFT GAPS FOR UNDISCOVERED ELEMENTS

EACH ROW HAS SIMILAR PROPERTIES

VALENCE 4

HE SWITCHED Te AND I TO MAKE VALENCES WORK!

```
                                    Ti=50    Zr=90    ?=180.
                                    V=51     Nb=94    Ta=182.
                                    Cr=52    Mo=96    W=186.
                                    Mn=55    Rh=104,4 Pt=197,1
                                    Fe=56    Ru=104,4 Ir=198.
                                  Ni=Co=59   Pl=106,6 Os=199.
                H=1                 Cu=63,4   Ag=108   Hg=200.
            Be=9,4  Mg=24   Zn=65,2            Cd=112
            B=11    Al=27,4   ?=68            Ur=116   Au=197?
            C=12    Si=28     ?=70            Sn=118
            N=14    P=31     As=75            Sb=122   Bi=210
            O=16    S=32     Se=79,4          Te=128?
            F=19    Cl=35,5  Br=80             I=127
       Li=7 Na=23   K=39     Rb=85,4          Cs=133   Tl=204
                    Ca=40    Sr=57,6          Ba=137   Pb=207.
                     ?=45    Ce=92
                   ?Er=56    La=94
                   ?Yt=60    Di=95
                   ?In=75,6  Th=118?
```

	EKA-ALUMINUM (MENDELEEV'S PREDICTION)	REAL GALLIUM
ATOMIC WEIGHT	68	69.9
DENSITY	6.0 g/cm^3	5.96 g/cm^3
ATOMIC VOLUME	11.5	11.7
MELTING POINT	LOW	30°C
OXIDE FORMULA	Ea_2O_3	Ga_2O_3
CHLORIDE FORMULA	Ea_2Cl_6	Ga_2Cl_6

WHOA!

NEXT, LARS FREDRICK NILSON (SWEDISH, 1840-99) FOUND AN ELEMENT THAT MATCHED MENDELEEV'S PREDICTION FOR EKA-BORON. NILSON NAMED HIS METAL *SCANDIUM* (FOR "SCANDINAVIA")!

	EKA-BORON (MENDELEEV'S PREDICTION)	REAL SCANDIUM
ATOMIC WEIGHT	44	43.79
OXIDE FORMULA	Eb_2O_3	Sc_2O_3
OXIDE DENSITY	3.5 g/cm^3	3.86 g/cm^3
SULFATE FORMULA	$Eb_2(SO_4)_3$	$Sc_2(SO_4)_3$

THIS IS GETTIN' *WEIRD!*

FOR THE *PIÈCE DE RÉSISTANCE*, IN 1886 CLEMENS ALEXANDER WINKLER (GERMAN, 1838-1904) DISCOVERED AN ELEMENT IN SILVER ORE. TO GET BACK AT THE FRENCH AND THEIR GALLIUM (AND MAYBE THE AUTHOR OF THIS BOOK AND HIS FRENCH EXPRESSIONS), WINKLER CALLED THE ELEMENT *GERMANIUM!*

	EKA-SILICON (MENDELEEV'S PREDICTION)	REAL GERMANIUM
ATOMIC WEIGHT	72	72.6
DENSITY	5.5 g/cm^3	5.47 g/cm^3
MELTING POINT	HIGH	958°C
OXIDE FORMULA	EsO_2	GeO_2
CHLORIDE FORMULA	$EsCl_4$	$GeCl_4$
SULFIDE FORMULA	EsS_2	GeS_2

INCONCEIVABLE!

NOW PEOPLE BELIEVED MENDELEEV!

ARGON'S ATOMIC WEIGHT IS 40... SO WHERE DOES IT GO ON THE PERIODIC TABLE?

WITH VALENCE 0, PUT IT HERE!

ELEMENT	AT. WT.	VALENCE
P	31	3
S	32	2
Cl	35.5	1
K	39	1
Ca	40	2
Sc	44	3

Ar

STRANGE BUT TRUE!

ENGLISH AUTHOR H.G. WELLS (1866–1946) WAS INTRIGUED WITH ARGON'S DISCOVERY! IN HIS BOOK *WAR OF THE WORLDS*...

LIKE WOW, MAN!

...THE MARTIANS USE A BLACK POWDER THAT COMBINES WITH ARGON TO KILL EARTH'S LIFE!

BUT ARGON IS *INERT*!

IF WE HAVE A NEW ROW OF INERT ELEMENTS IN OUR PERIODIC TABLE, WHERE ARE ALL THE OTHERS IN THAT ROW?

RAMSAY BEGAN LOOKING...

HE BOUGHT SAMPLES OF CLEVEITE, A URANIUM-CONTAINING MINERAL KNOWN TO GIVE OFF GAS!

HE CRUSHED IT AND BOILED IT IN SULFURIC ACID, AND EXTRACTED THE GAS...

PUMP

GAS

HEAT

ACID WITH CLEVEITE BITS

...AND— INTO THE SPECTROSCOPE IT WENT...

ARGON

NEW GAS

WE'VE GOT A NEW ELEMENT!

IT MATCHED THE SPECTRUM FROM A SOLAR ECLIPSE IN 1869, WHEN THE ELEMENT WAS CALLED *HELIUM* (GREEK ἥλιος, HELIOS, "SUN")!

HELIUM WAS FIRST FOUND ON THE SUN!

Atomic weight of Helium is 4

Valence = 0

CAN YOU GUESS WHERE HELIUM GOES IN THE TABLE?

ELEMENT	AT. WT.	VALENCE
H	1.01	1
Li	7	1
Be	9	2
B	11	3

He →

HELIUM FITS PERFECTLY BETWEEN HYDROGEN AND LITHIUM... BUT ARE THERE *OTHER* VALENCE 0 ELEMENTS?

IN 1898, RAMSAY DISCOVERED THREE MORE GASES WITH VALENCE 0, BY BOILING OFF LIQUID AIR AND TRAPPING THE RESIDUES...

NEON (GREEK νέον, "NEW")

KRYPTON (GREEK κρυπτός, "HIDDEN ONE")

XENON (GREEK ξένον, "STRANGE")

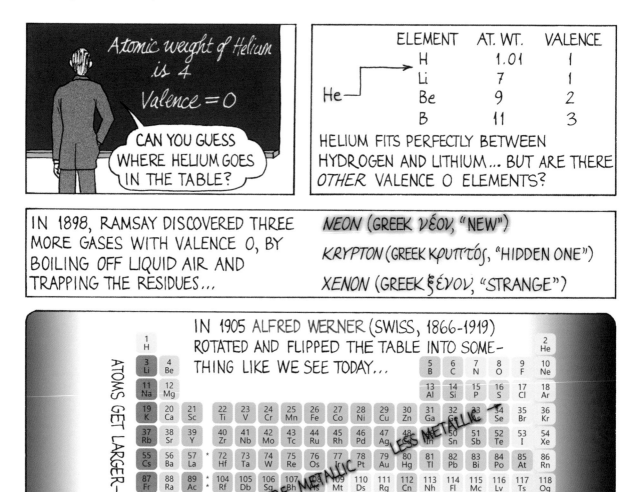

IN 1905 ALFRED WERNER (SWISS, 1866-1919) ROTATED AND FLIPPED THE TABLE INTO SOMETHING LIKE WE SEE TODAY...

ATOMS GET LARGER →

← ATOMS GET LARGER

MORE METALLIC

LESS METALLIC

PRIMO LEVI (ITALIAN, 1919-1987), A JEWISH CHEMIST WHO WAS IMPRISONED BY THE NAZIS, PUBLISHED HIS *IL SISTEMA PERIODICO**, WITH 21 CHAPTERS TITLED AFTER ELEMENTS, DESCRIBING SCENES FROM HIS LIFE!

ALSO CHECK OUT NOTTINGHAM UNIVERSITY'S PERIODIC TABLE VIDEOS AT PERIODICVIDEOS.COM!

*ENGLISH TITLE: *THE PERIODIC TABLE*

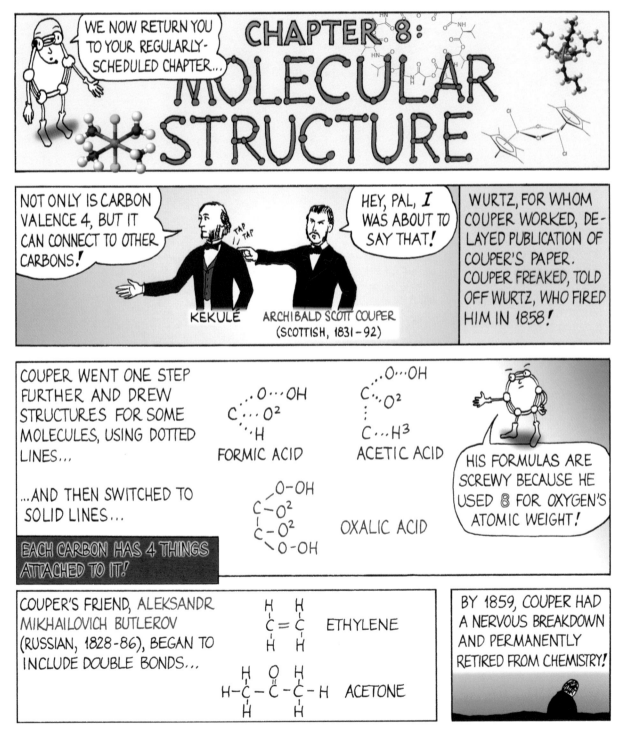

CHAPTER 8: MOLECULAR STRUCTURE

WE NOW RETURN YOU TO YOUR REGULARLY-SCHEDULED CHAPTER...

NOT ONLY IS CARBON VALENCE 4, BUT IT CAN CONNECT TO OTHER CARBONS!

HEY, PAL, *I* WAS ABOUT TO SAY THAT!

KEKULÉ

ARCHIBALD SCOTT COUPER
(SCOTTISH, 1831-92)

WURTZ, FOR WHOM COUPER WORKED, DELAYED PUBLICATION OF COUPER'S PAPER. COUPER FREAKED, TOLD OFF WURTZ, WHO FIRED HIM IN 1858!

COUPER WENT ONE STEP FURTHER AND DREW STRUCTURES FOR SOME MOLECULES, USING DOTTED LINES...

FORMIC ACID

ACETIC ACID

...AND THEN SWITCHED TO SOLID LINES...

OXALIC ACID

EACH CARBON HAS 4 THINGS ATTACHED TO IT!

HIS FORMULAS ARE SCREWY BECAUSE HE USED 8 FOR OXYGEN'S ATOMIC WEIGHT!

COUPER'S FRIEND, ALEKSANDR MIKHAILOVICH BUTLEROV (RUSSIAN, 1828-86), BEGAN TO INCLUDE DOUBLE BONDS...

ETHYLENE

ACETONE

BY 1859, COUPER HAD A NERVOUS BREAKDOWN AND PERMANENTLY RETIRED FROM CHEMISTRY!

79

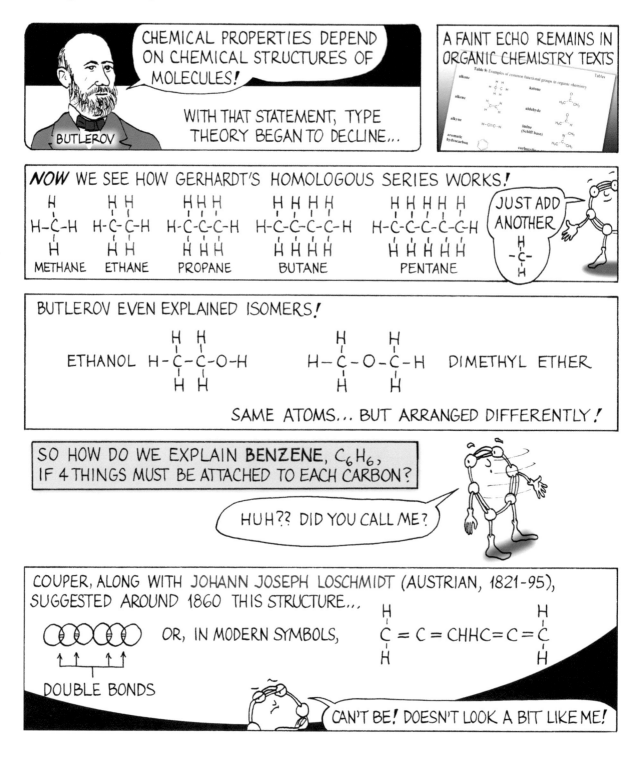

CHEMICAL PROPERTIES DEPEND ON CHEMICAL STRUCTURES OF MOLECULES!

BUTLEROV

WITH THAT STATEMENT, TYPE THEORY BEGAN TO DECLINE...

A FAINT ECHO REMAINS IN ORGANIC CHEMISTRY TEXTS

NOW WE SEE HOW GERHARDT'S HOMOLOGOUS SERIES WORKS!

METHANE   ETHANE   PROPANE   BUTANE   PENTANE

JUST ADD ANOTHER
H
–C–
H

BUTLEROV EVEN EXPLAINED ISOMERS!

ETHANOL

DIMETHYL ETHER

SAME ATOMS... BUT ARRANGED DIFFERENTLY!

SO HOW DO WE EXPLAIN **BENZENE**, C_6H_6, IF 4 THINGS MUST BE ATTACHED TO EACH CARBON?

HUH?? DID YOU CALL ME?

COUPER, ALONG WITH JOHANN JOSEPH LOSCHMIDT (AUSTRIAN, 1821-95), SUGGESTED AROUND 1860 THIS STRUCTURE...

OR, IN MODERN SYMBOLS,

$C=C=CHHC=C=C$

DOUBLE BONDS

CAN'T BE! DOESN'T LOOK A BIT LIKE ME!

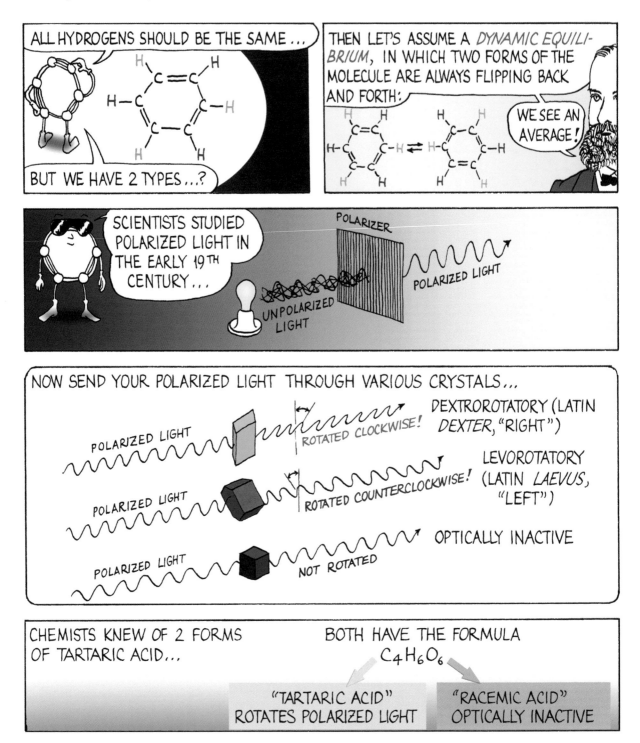

REMEMBER KEKULÉ'S INSIGHT THAT CARBON ATOMS CAN FORM RINGS?

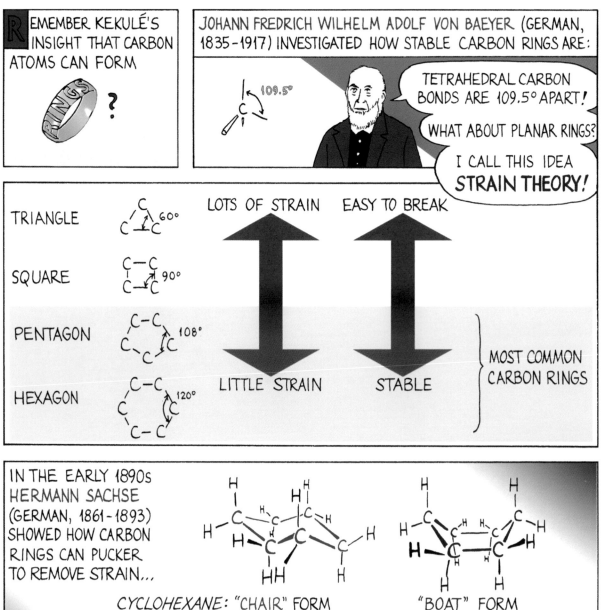

JOHANN FREDRICH WILHELM ADOLF VON BAEYER (GERMAN, 1835–1917) INVESTIGATED HOW STABLE CARBON RINGS ARE:

109.5°

TETRAHEDRAL CARBON BONDS ARE 109.5° APART!

WHAT ABOUT PLANAR RINGS?

I CALL THIS IDEA **STRAIN THEORY!**

TRIANGLE   60°

SQUARE   90°

PENTAGON   108°

HEXAGON   120°

LOTS OF STRAIN   EASY TO BREAK

LITTLE STRAIN   STABLE

} MOST COMMON CARBON RINGS

IN THE EARLY 1890s HERMANN SACHSE (GERMAN, 1861–1893) SHOWED HOW CARBON RINGS CAN PUCKER TO REMOVE STRAIN...

CYCLOHEXANE: "CHAIR" FORM

"BOAT" FORM

NOBODY BELIEVED SACHSE TILL EXPERIMENTAL TECHNIQUES CAUGHT UP 25 YEARS LATER...

BENZENE, WITH ITS WEIRD ALTERNATING SINGLE AND DOUBLE BONDS, DOES NOT PUCKER, AND STAYS FLAT.

CHEMISTS CAN BE LAZY (LIKE EVERYONE ELSE), SO THEY MAKE THEIR DRAWINGS OF MOLECULES EASIER AND SIMPLER:

BECAUSE CARBON ALMOST ALWAYS HAS A VALENCE OF 4, WE SKIP DRAWING THE HYDROGENS, AND ASSUME EXTRA HYDROGEN ATOMS FILL UP THE 4 SPOTS AROUND EACH CARBON ATOM...

...AND EACH VERTEX WHERE LINES MEET IS A CARBON ATOM...

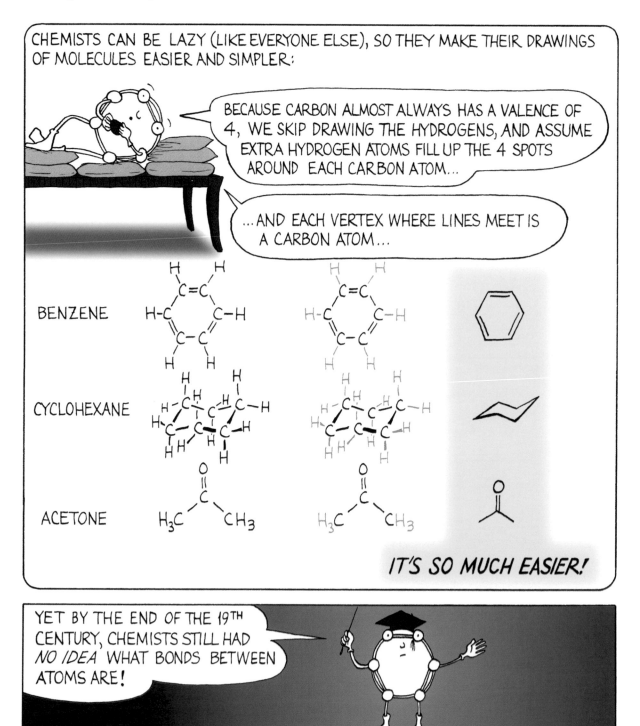

BENZENE

CYCLOHEXANE

ACETONE

IT'S SO MUCH EASIER!

YET BY THE END OF THE 19TH CENTURY, CHEMISTS STILL HAD *NO IDEA* WHAT BONDS BETWEEN ATOMS ARE!

CHAPTER 9:
PHYSICAL CHEMISTRY

LET'S GET PHYSICAL.... PHYSICAL....♪

LONDON, 1808: A TEENAGE BOY...

MICHAEL FARADAY (1791 - 1867) WAS A BOOKBINDER'S APPRENTICE...

WITH PERMISSION, HE READ SOME OF THE BOOKS...

...INCLUDING JANE MARCET'S *CONVERSATIONS ON CHEMISTRY!*

HE GOT A TICKET IN 1812 TO ONE OF HUMPHRY DAVY'S FAMOUS LECTURES...

...TOOK CAREFUL NOTES...

...AND APPLIED TO BE DAVY'S ASSISTANT!

FOUR LECTURES
being part of a Course on
The Elements of
CHEMICAL PHILOSOPHY
Delivered by
SIR H. DAVY
LLD ScRS FRSE MRIA MRI &c.
at the
Royal Institution.
And taken of from Notes
BY
M. FARADAY
1812

DAVY ACCEPTED HIM IN 1813!

BY 1825, FARADAY WAS DIRECTOR OF THE LABORATORY AT THE ROYAL INSTITUTION...

...WHEN HE DISCOVERED *BENZENE!*

FARADAY'S LABORATORY AT THE ROYAL INSTITUTION

MY ANCESTOR? IN A *BOTTLE??*

WITH A QUANTITATIVE UNDERSTANDING, ELECTROCHEMISTRY BECAME COMMERCIAL...

DANIELL CELL

PLANTÉ'S LEAD-ACID CELL (USED IN CARS)

LECLANCHÉ CELL (ITS DESCENDANTS ARE ALKALINE CELLS)

A VARIETY OF BATTERIES

ENGLAND, 1840S

RUSSIAN RELIGIOUS ICON

ELECTROPLATING CHEAP METAL OBJECTS

BUT HOW DID IT WORK? CHEMISTS ARGUED...

UM... IONS CARRY CURRENT IN A SOLUTION...

THE CHARGES COME SOMEHOW FROM WIRES...

BUT WHAT *IS* AN ION? HOW DOES IT CARRY CURRENT?

THESE JOKERS HAVE NO CLUE... LET'S MOVE ON...

BY THE 1840s, SCIENTISTS KNEW A BIT ABOUT THE FLOW OF HEAT, OR *THERMODYNAMICS* — FOR EXAMPLE:

ENERGY IS NEITHER CREATED NOR DESTROYED!
— FIRST LAW OF THERMODYNAMICS

HEAT FLOWS SPONTANEOUSLY FROM

HIGH TEMPERATURE
TO
LOW TEMPERATURE

CHEMICAL REACTIONS MAKE HEAT*!*

BAKING SODA

VINEGAR

WHILE HEAT FLOWS, YOU CAN DO WORK...

NO*!* NOT TILL 1942*!*

ALL CHEMICAL REACTIONS TRANSFER HEAT*!*

SYSTEM (THE REACTION)

HEAT CAN LEAVE OR ENTER

SURROUNDINGS

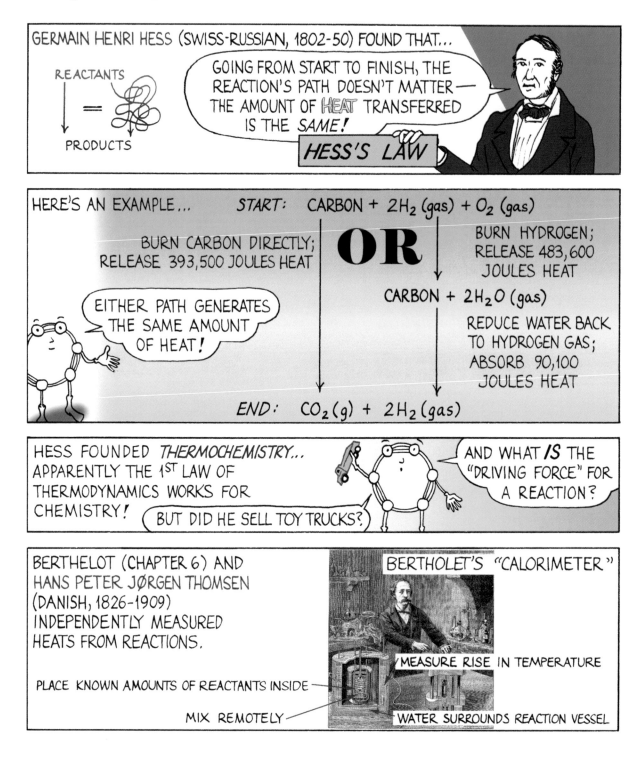

A Tale of Political Intrigue

THE TOKUGAWA SHOGUNATE RAN JAPAN...

CHŌSHŪ HAN — TOKUGAWA SHOGUNATE

A RIVAL DOMAIN, THE CHŌSHŪ HAN, WANTED TO GAIN WESTERN KNOWLEDGE AND TECHNOLOGY...

BUT IT WAS *ILLEGAL* IN THE EARLY 1860s TO LEAVE

JAPAN 日本

THE CHŌSHŪ HAN SMUGGLED FIVE COLLEGE STUDENTS, THE *CHŌSHŪ FIVE*, TO LONDON IN 1863!

PROFESSOR WILLIAMSON AND HIS WIFE CATHERINE WERE IN CHARGE OF THE *CHŌSHŪ FIVE* AT UNIVERSITY COLLEGE LONDON!

IN NORWAY, CATO MAXIMILIAN GULDBERG (1836-1902) AND HIS BROTHER-IN-LAW PETER WAAGE (1833-1900) STUDIED SPONTANEOUS REACTIONS, AND BEGAN PUBLISHING RESEARCH IN 1864...

$$A + B \underset{\text{REVERSE}}{\overset{\text{FORWARD}}{\rightleftharpoons}} C + D$$

THE POINT OF EQUILIBRIUM DEPENDS ON RELATIVE RATES! IF **FORWARD** RATE IS GREATER THAN **REVERSE** RATE, THE A+B → C+D REACTION IS FAST, AND THE C+D → A+B RATE IS SLOW— SO WE SEE AN EQUILIBRIUM OF MOSTLY C+D!

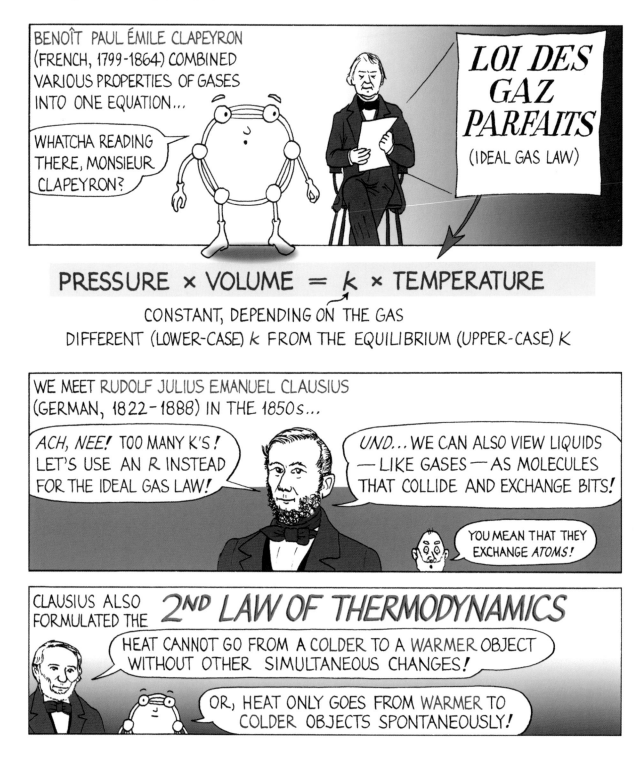

BENOÎT PAUL ÉMILE CLAPEYRON (FRENCH, 1799-1864) COMBINED VARIOUS PROPERTIES OF GASES INTO ONE EQUATION...

WHATCHA READING THERE, MONSIEUR CLAPEYRON?

LOI DES GAZ PARFAITS (IDEAL GAS LAW)

PRESSURE × VOLUME = k × TEMPERATURE

CONSTANT, DEPENDING ON THE GAS
DIFFERENT (LOWER-CASE) k FROM THE EQUILIBRIUM (UPPER-CASE) K

WE MEET RUDOLF JULIUS EMANUEL CLAUSIUS (GERMAN, 1822-1888) IN THE 1850s...

ACH, NEE! TOO MANY K'S! LET'S USE AN R INSTEAD FOR THE IDEAL GAS LAW!

UND...WE CAN ALSO VIEW LIQUIDS —LIKE GASES— AS MOLECULES THAT COLLIDE AND EXCHANGE BITS!

YOU MEAN THAT THEY EXCHANGE *ATOMS!*

CLAUSIUS ALSO FORMULATED THE *2ND LAW OF THERMODYNAMICS*

HEAT CANNOT GO FROM A COLDER TO A WARMER OBJECT WITHOUT OTHER SIMULTANEOUS CHANGES!

OR, HEAT ONLY GOES FROM WARMER TO COLDER OBJECTS SPONTANEOUSLY!

TWO SCIENTISTS — JAMES CLERK MAXWELL (SCOTTISH, (1831-79) AND LUDWIG EDUARD BOLTZMANN (AUSTRIAN, 1844-1906) — ASSUMED GASES ARE COLLECTIONS OF RANDOMLY MOVING PARTICLES...

THE IDEAL GAS LAW IS VALID ONLY IF GAS PARTICLES DON'T ATTRACT...

...AND GAS PARTICLES HAVE NO SIZE!

MAXWELL

BOLTZMANN

BUT THIS *KINETIC THEORY OF GASES* MUST DEAL WITH THE FACT THAT GAS MOLECULES *DO* ATTRACT AND HAVE VOLUME!

A GAS

SO IN 1873 JOHANNES DIDERIK VAN DER WAALS (DUTCH, 1837-1923) MODIFIED THE IDEAL GAS LAW...

$$\left(PRESSURE + \frac{MOLECULAR\ ATTRACTION}{} \right) \times \left(VOLUME - \frac{MOLECULAR\ VOLUME}{} \right) = R \times TEMPERATURE$$

WORKS A LOT BETTER!

CLAUSIUS CONTINUED WITH SPONTANEOUS REACTIONS. HE INVENTED THE IDEA OF

ENTROPY*

OR "TRANSFORMATIONAL CONTENT" OR EVEN "AMOUNT OF DISORDER"

*GREEK ἐν (EN, IN) + τροπή (TROPE, TRANSFORMATION)

$$\frac{HEAT\ CONTENT\ OF\ ISOLATED\ SYSTEM}{ABSOLUTE\ TEMPERATURE} = ENTROPY$$

ENTROPY ALWAYS *INCREASES* DURING A SPONTANEOUS REACTION!

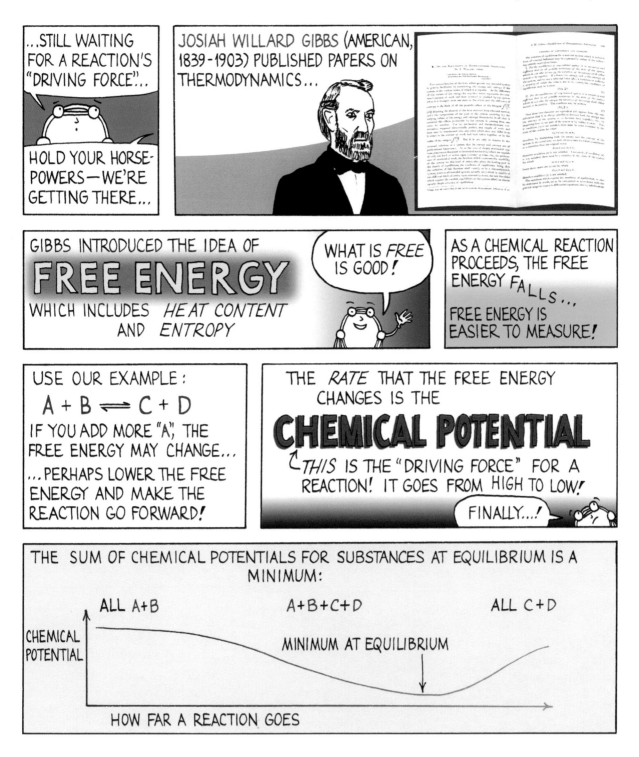

EQUILIBRIA EXIST FOR TRANSITIONS FROM

SOLID TO LIQUID   LIQUID TO GAS   "DRY ICE" (SOLID CO_2)   SOLID TO GAS

ICE, WATER, AND WATER VAPOR CAN CO-EXIST AT A "TRIPLE POINT"...

...A PRESSURE OF 0.006 ATMOSPHERES AND A TEMPERATURE OF 0.01°C!

GIBBS MATHEMATICALLY DERIVED THE

PHASE RULE

TO PREDICT HOW TEMPERATURE, PRESSURE, AND CONCENTRATION ARE RELATED...

...BUT HE PUBLISHED IN AN OBSCURE *AMERICAN* JOURNAL WHICH EUROPEANS IGNORED!

DUMMER AMI!

LES STUPIDES AMÉRICAINS BARBARES!

HOW UNCIVILISED OF HIM!

SO LET'S LOOK AT **CATALYSIS** INSTEAD

IN THE EARLY 1800s, CHEMISTS SAW TWO CAUSES FOR CHEMICAL REACTIONS:

1. MOLECULES ARE ALWAYS MOVING AND BANGING INTO EACH OTHER
2. CHEMICAL *"AFFINITIES"* AMONG ELEMENTS

WHATEVER AFFINITY IS...

HEATING CAUSES MORE MOLECULAR MOTION — AND MORE REACTIONS!

BUT WHAT ABOUT *FERMENTATION* — THE ACTION OF YEAST?

BREWING BEER

DOUGH RISING

SO IF HEATING REACTANTS DOESN'T WORK, CAN I ADD A FERMENTING AGENT — LIKE A PHILOSOPHER'S STONE?

CHEMISTS KNEW THAT PLACING A HOT PLATINUM WIRE OR POWDER IN ALCOHOL LIBERATES OXYGEN AND GLOWS!

GLOWING Pt WIRE

ALCOHOL VAPOR
ALCOHOL

LIKEWISE, PLANT FIBERS, ANIMAL TISSUES, AND METALS ALL BREAK DOWN HYDROGEN PEROXIDE...

$$2H_2O_2 \xrightarrow[\text{PLANT FIBERS}]{\text{WITH}} 2H_2O + O_2$$

SVANTE AUGUST ARRHENIUS (SWEDISH, 1859 – 1927) WAS STUDYING ELECTROLYTE SOLUTIONS FOR HIS DOCTORATE:

ELECTROLYTE SOL-UTIONS CARRY ELECTRICITY!

FARADAY

ARRHENIUS PRESENTED HIS DISSERTATION IN 1884:

ELECTROLYTES LIKE NaCl AND CaCl$_2$ *DISSOCIATE* (BREAK UP) INTO ATOMS THAT CARRY CURRENT! *THEY* ARE THE IONS!

HMMPH... WE'LL LET YOU SLIDE BY!

HE USED TABLE SALT AS AN EXAMPLE...

Na$^+$ Na$^+$ Cl$^-$ Na$^+$
Cl$^-$ Cl$^-$ Na$^+$ Cl$^-$
Na$^+$ Na$^+$ Cl$^-$ Cl$^-$
Cl$^-$ Na$^+$ Cl$^-$ Na$^+$ Na$^+$
Cl$^-$ Cl$^-$

♪ BREAKING UP IS HARD TO DO! ♪
B

FARADAY'S "IONS" ARE ATOMS (OR GROUPS) WITH A ⊕ OR ⊖ CHARGE!

WHY WERE HIS PROFESSORS NOT THRILLED?
① ARRHENIUS COULDN'T EXPLAIN HOW POISONOUS CHLORINE ATOMS IN SOLUTION WERE SAFE TO DRINK
② HE COULDN'T EXPLAIN WHY Na ATOMS DIDN'T BURST INTO FLAME IN WATER
③ HE COULDN'T EXPLAIN WHERE THE ELECTRIC CHARGES CAME FROM

BUT HE WAS RIGHT! IONIC DISSOCIATION EXPLAINED FREEZING POINT DEPRESSION NICELY...

TABLE SUGAR (SUCROSE) DOES *NOT* DISSOCIATE
NaCl $\longrightarrow$ Na$^+$ + Cl$^-$ WITH *TWICE* THE PARTICLES
CaCl$_2$ $\longrightarrow$ Ca^{++} + Cl$^-$ + Cl$^-$ WITH *THREE TIMES* THE PARTICLES!

BY 1889, ARRHENIUS NOTED THAT MOLECULES DON'T ALWAYS REACT IN A COLLISION... RATHER, THEY NEED A MINIMUM ENERGY, AN *ACTIVATION ENERGY*, TO START:

FREE ENERGY

REACTANTS

↑ LARGE ACTIVATION ENERGY

CHANGE IN FREE ENERGY

PRODUCTS

PROGRESS OF **DIFFICULT** REACTION

FREE ENERGY

REACTANTS

SMALL ACTIVATION ENERGY

CHANGE IN FREE ENERGY

PRODUCTS

PROGRESS OF **EASY** REACTION

WHEN YOU RAISE THE TEMPERATURE OF A REACTANT, YOU GIVE IT MORE ENERGY TO REACT FASTER—TO THE POINT OF IGNITION!

$$2H_2 + O_2 \rightarrow 2H_2O$$

AUTOIGNITION AT 536 °C

MANY MOLECULES NEED AN EXTRA KICK TO REACT...

SOME MORE THAN OTHERS!

WE RETURN TO CATALYSIS...

FRIEDRICH WILHELM OSTWALD (BALTIC GERMAN, 1853-1932) FINALLY TRANSLATED GIBBS'S WORK INTO GERMAN IN 1892...

GIBBS IS AMAZING! THIS IS EXACTLY WHAT WE NEED TO UNDERSTAND CATALYSIS!

REACTION RATE SPEEDS UP WITH A CATALYST!

...AND THE CATALYST IS *NOT* USED UP DURING THE REACTION!

THE FREE-ENERGY CHANGE IS THE *SAME* WITH OR WITHOUT THE CATALYST

THE REACTANTS TEMPORARILY COMBINE WITH THE CATALYST TO MAKE AN "INTERMEDIATE" (WITH LOWER ACTIVATION ENERGY). THE INTERMEDIATE RELEASES THE PRODUCTS!

FREE ENERGY

NO CATALYST

REACTANTS

INTERMEDIATE

PRODUCTS

PROGRESS OF THE REACTION

THIS ALSO EXPLAINS *ENZYMES** (PROTEIN-BASED CATALYSTS) INSIDE LIVING CREATURES!

* GREEK ἐν (*EN*, "IN") + ζύμη (*ZÚMĒ*, "SOURDOUGH")

REMEMBER HESS'S LAW? A CATALYST OFFERS AN ALTERNATE PATHWAY TO THE PRODUCTS WITH A LOWER ACTIVATION ENERGY!

WITH ALL THESE PHYSICAL EXPLANATIONS COMING FAST, ARRHENIUS, VAN'T HOFF, AND OSTWALD FOUNDED *ZEITSCHRIFT FÜR PHYSIKALISCHE CHEMIE* IN 1887...

SO IT'S AGREED!

CAN WE HAVE A SECRET HANDSHAKE?

IT STILL EXISTS TODAY!

THE OFFICIAL START OF PHYSICAL CHEMISTRY!

HENRY-LOUIS LE CHÂTELIER (FRENCH, 1850-1936) TRANSLATED GIBBS'S WORK INTO FRENCH IN 1899. HE USED GIBBS'S IDEAS...

A CHANGE IN ONE FACTOR OF EQUILIBRIUM REARRANGES THE SYSTEM TO MINIMIZE THAT CHANGE!

LE CHÂTELIER'S PRINCIPLE

FOR EXAMPLE, MAKING AMMONIA:
$$N_2(g) + 3H_2(g) \rightleftharpoons 2NH_3(g) + 92000\,J$$
(HEAT)

IF WE SQUEEZE THE SYSTEM—WITH 4 UNITS OF GAS ON THE LEFT, BUT 2 ON THE RIGHT—WE GET MORE PRODUCT (PLUS HEAT)!

IF WE *HEAT* THE SYSTEM, THERE IS LESS HEAT ON THE LEFT, SO THE SYSTEM SHIFTS TO MORE REACTANTS!

WALTHER HERMANN NERNST (GERMAN, 1864-1941) APPLIED THERMODYNAMICS TO ELECTROCHEMISTRY AND BATTERIES...

$$E_{cell} = E^{\ominus}_{cell} - \frac{RT}{zF} \ln Q_r$$

NERNST LATER CAME UP WITH THE *THIRD* LAW OF THERMODYNAMICS:

YOU *CAN'T* GET DOWN TO ABSOLUTE ZERO IN A FINITE NUMBER OF STEPS! *SCHADE!*

Musical fun fact! AROUND 1930 NERNST HELPED CREATE THE *NEO-BECHSTEIN*, AN ELECTRONIC GRAND PIANO!

LET'S NOW CONSIDER LIGHT AND CHEMISTRY...

THIS SPACE INTENTIONALLY LEFT BLANK

PHOTOCHEMISTRY: LIGHT CAN CAUSE CHEMICAL REACTIONS!

Don't Panic!

LET'S EAT GRANDMA LET'S EAT, GRANDMA — COMMAS SAVE LIVES!

FULHAME'S WORK IN THE 1790s EVENTUALLY LED TO *PHOTOGRAPHY** IN THE 1830s...

(BUT NO SELFIES TILL LATER)

*GREEK φῶς (*PHOS*, "LIGHT" + γραφή (*GRAPHĒ*, "DRAWING"))

TINY SILVER HALIDE CRYSTALS EMBEDDED IN GELATIN

PHOTO PLATE

PLATE

LIGHT INCREASES THE TENDENCY OF SILVER HALIDE TO BREAK DOWN INTO SILVER METAL

AFTER EXPOSURE, DIP PLATE OR FILM INTO "DEVELOPER" SOLUTION TO PERFORM BREAKDOWN OF SILVER HALIDE...

FIRST PHOTO OF A PERSON WAS TAKEN BY

LOUIS JACQUES MANDÉ DAGUERRE IN 1838!

FIRST COLOR PHOTO WAS TAKEN BY

HE USED JAMES CLERK MAXWELL'S THEORY OF THREE-COLOR VISION!

THOMAS SUTTON IN DECEMBER 1860!

LIGHT CAN EVEN BE A CATALYST! NERNST EXPLAINED THE MECHANISM OF

$$H_2(g) + Cl_2(g) \xrightarrow{\text{LIGHT}} 2HCl$$

1   H_2 GAS Cl_2 GAS

2   H_2 GAS Cl_2 GAS

3   BOOM!

WHY?

ACH! IT GOES LIKE THIS...

LIGHT SPLITS A CHLORINE— CHLORINE BOND...

SINGLE CHLORINE ATOMS ARE *VERY* REACTIVE, AND BREAK HYDROGEN—HYDROGEN BONDS:

$$Cl + H_2 \longrightarrow HCl + H$$
$$Cl + H_2 \longrightarrow HCl + H$$

(2 EQUATIONS BECAUSE 2 CHLORINES ARE MADE)

SINGLE HYDROGEN ATOMS ARE *ALSO* REACTIVE, AND ATTACK CHLORINE BONDS...

$$H + Cl_2 \longrightarrow HCl + Cl$$
$$H + Cl_2 \longrightarrow HCl + Cl$$

} THESE CHLORINES ATTACK MORE HYDROGENS...

...AND WE HAVE A **CHAIN REACTION** IN THE BLINK OF AN EYE!

IN 1828, BOTANIST ROBERT BROWN (SCOTTISH, 1773–1858) REPORTED THAT POLLEN GRAINS VIBRATED ENDLESSLY WHEN SUSPENDED IN A LIQUID...

WHAT *IS* IT? PERPETUAL MOTION?

ALBERT EINSTEIN (GERMAN-SWISS-AMERICAN, 1879-1955) SHOWED IN 1905 THAT THE LIQUID MOLECULES BOUNCING AGAINST THE POLLEN DID THIS!

HE GAVE AN EQUATION TO CALCULATE THE SIZE OF MOLECULES!

JEAN-BAPTISTE PERRIN (FRENCH, 1870-1942) DID THESE CALCULATIONS...

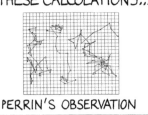

PERRIN'S OBSERVATION

PHYSICISTS FINALLY ACCEPTED THAT ATOMS & MOLECULES ARE REAL!

LET'S *PAR-TAY!*

BUT NOW LET'S RETURN TO EQUILIBRIUM... OF WATER!

GLUG GLUG GLUG

WATER PARTIALLY DISSOCIATES SPONTANEOUSLY:

$$H_2O \rightleftharpoons H^+ + OH^-$$

THE EQUILIBRIUM CONSTANT, K, FOR WATER IS

$$K = \frac{H^+ \text{ CONCENTRATION} \times OH^- \text{ CONCENTRATION}}{H_2O \text{ CONCENTRATION}}$$

(TECHNICALLY H^+ DOESN'T EXIST IN WATER — IT'S REALLY MORE LIKE H_3O^+, BUT THE MODEL STILL WORKS WELL, SO PLAY ALONG...)

ONLY A *TEEENSY* FRACTION OF THE WATER BREAKS APART INTO IONS! SO LET'S PRETEND THAT THE WATER CONCENTRATION DOESN'T CHANGE DURING DISSOCIATION...

WE REMOVE THE "H_2O CONCENTRATION" FROM THE EQUATION:

$$K = H^+ \text{ CONCENTRATION} \times OH^- \text{ CONCENTRATION}$$

IN PURE WATER, ONLY 1 IN 10 MILLION MOLECULES BREAKS UP INTO EQUAL AMOUNTS OF H^+ AND OH^-

SO H^+ CONCENTRATION = OH^- CONCENTRATION = 10^{-7}

WHICH MEANS $K = 10^{-7} \times 10^{-7} = 10^{-14}$!

SØREN PETER LAURITZ SØRENSEN (DANISH, 1868-1939), CHIEF CHEMIST AT THE CARLSBERG BREWERY, HAD SOME IDEAS ABOUT THIS IN 1909...

Y'KNOW, IF WE REMOVED THE EXPONENT, IT WOULD BE EASIER!

AND SO SØRENSON REMOVED THE EXPONENT *AND* THE NEGATIVE SIGN...

I CALL IT THE pH SCALE: IT'S AN EASY SCALE TO SHOW THE H⁺ CONCENTRATION*!*

H^+ IS ACIDIC AND OH^- IS BASIC, SO

ALL ACID	pH = 0
NEUTRAL	pH = 7
ALL BASE	pH = 14

LEONOR MICHAELIS (GERMAN, 1875–1949) WAS WORKING ON THE SAME IDEA BUT SØRENSON PUBLISHED FIRST*!*

MICHAELIS POPULARIZED THE pH SCALE IN HIS 1914 BOOK...

I GAVE MY BOOK THE IMPRESSIVE TITLE *DIE WASSERSTOFFIONENKONZENTRATION!* *

YIKES*!* THAT'S 29 LETTERS IN ONE WORD*!*

* "THE CONCENTRATION OF HYDROGEN IONS"

ALL THIS CHEMICAL KNOWLEDGE EVEN FILTERED DOWN TO KIDS...

LUCY RIDER MEYER'S BOOK FROM 1887, *REAL FAIRY FOLKS, OR FAIRY LAND OF CHEMISTRY: EXPLORATIONS IN THE WORLD OF ATOMS*

SOME OF THE REAL FAIRY FOLKS. *Frontis*

FAIRIES OF THE AIR.

FAIRY PICTURE OF WATER.

MEYERS GOT THE PROPORTION OF NITROGEN TO OXYGEN IN THE AIR CORRECT: 4 TO 1 *!*

≥*PANT!*≤ I AM EXHAUSTED*!* 21 PAGES IN THIS CHAPTER*!* ≥*PANT!*≤ ENOUGH ALREADY WITH GETTING PHYSICAL*!*

CHAPTER 10! INDUSTRIAL CHEMISTRY

BLUE PIGMENTS WERE UNRELIABLE FOR ARTISTS: THEY WERE UNSTABLE OR RARE...

ULTRAMARINE FROM AFGHANISTAN MADE FROM LAPIS LAZULI 😊 😊 😊

INDIGO FADES

AZURITE TURNS GREEN IN WATER

I, JOHANN JACOB DIESBACH, ACCIDENTALLY INVENTED THE FIRST SYNTHETIC PIGMENT IN 1706!

COCHINEAL INSECT, CRUSHED

POTASH

IRON SULFATE IMPURITY

PRUSSIAN BLUE!

THE FIRST KNOWN USE OF PRUSSIAN BLUE IS IN PIETER VAN DE WERFF'S 1709 *ENTOMBMENT OF CHRIST!*

18TH CENTURY ART BECAME ALIVE WITH BLUE!

ANOTHER SYNTHETIC DYE FROM THE 18TH CENTURY IS *PICRIC ACID*, DISCOVERED BY PETER WOULFE (IRISH, 1727–1803), WHEN HE TREATED INDIGO WITH NITRIC ACID!

IT WAS GOOD FOR DYEING WOOL & SILK!

BESIDES THE FABRIC INDUSTRY, GAS ILLUMINATION BEGAN IN THE 1790s...

FACTORIES DISTILLED GAS FROM WOOD AND COAL!

JAMES MUSPRATT (1793–1886) FOUNDED THE BRITISH CHEMICAL INDUSTRY...

FIRST I MADE PRUSSIAN BLUE...

...THEN MORDANTS,* WHICH FIX DYE TO FABRICS...

...THEN BLEACHES FOR IRISH LINENS! ✽✽✽✽✽✽✽✽✽

I FABRICATED MY BUSINESS OUT OF WHOLE CLOTH! HAHA

*LATIN *MORDERE*, "TO BITE"

MUSPRATT ALSO MANUFACTURED *SODA* (SODIUM CARBONATE, Na_2CO_3), USED FOR...

GLASS

CLOTH

PAPER

SOAP

CAN CHEMISTS SYNTHESIZE IT?

PLUS, GROWING POPULATIONS MEANT MORE FERTILIZERS NEEDED IN AGRICULTURE...

... AND BETTER EXTRACTION OF SUGAR FROM A VARIETY OF SOURCES WAS NEEDED!

SWEET!

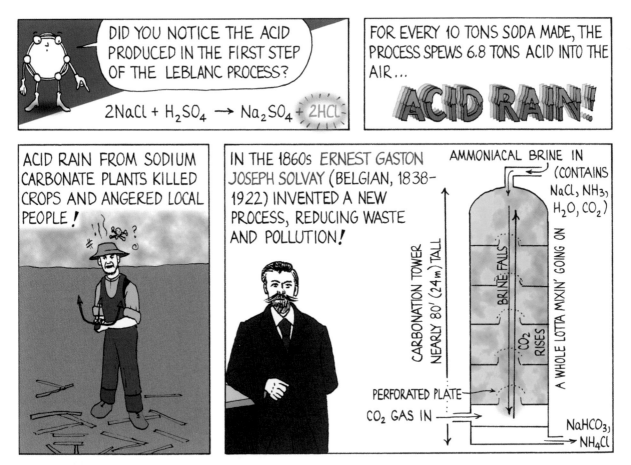

BRITAIN RECRUITED AUGUST WILHELM VON HOFMANN (GERMAN, 1818-1892) TO FOUND A COLLEGE OF CHEMISTRY IN LONDON IN 1845...

BESSEMER'S PROCESS RESULTED IN STEEL IN INDUSTRIAL QUANTITIES...

THE AIR BURNS OFF IMPURITIES IN THE IRON!

BESSEMER'S STEEL WAS USED FOR RAILS IN THE U.S.A....

MOLTEN PIG IRON

AIR

...AND RAILROADS SPROUTED THROUGHOUT THE UNITED STATES!

IN THE EARLY 1880s, ROBERT ABBOTT HADFIELD (ENGLISH, 1858–1940) DISCOVERED THAT 13% MANGANESE IN IRON, HEATED TO 1000°C, THEN QUENCHED IN WATER...

...BECAME A STEEL EVEN STRONGER THAN MY STARCHED COLLAR!

はじめまして。*
I, KOTARO HONDA (JAPAN, 1870–1954), FOUND THAT CHROMIUM ADDED TO TUNGSTEN STEEL MADE POWERFUL MAGNETS!

*HAJIMEMASHITE, "GREETINGS!"

AND IN 1912, ELWOOD HAYNES (AMERICAN, 1857–1925), WAS BUILDING NEW-FANGLED CARS...

THE HAYNES MODEL "Y"

I NEEDED CORROSION-RESISTANT CAR PARTS!

HAYNES PATENTED STEEL ALLOYS OF IRON, TUNGSTEN, AND CHROMIUM, INCLUDING STAINLESS STEEL!

WATER

WHAT ABOUT ALUMINUM?

ALUMINUM WAS SO HARD TO WIN FROM ITS ORE THAT IT BECAME A PRECIOUS METAL!

APEX OF WASHINGTON MONUMENT

FRENCH COIN

BRACELET

OTHER COUNTRIES BEGAN WHALING—WITH COMPETITION, CAN AMERICA FIND OTHER OIL?

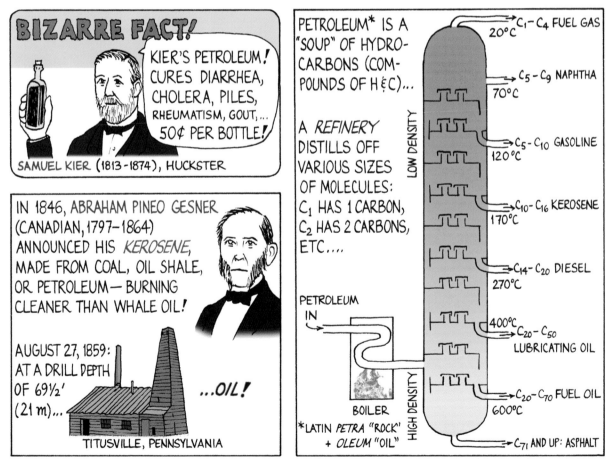

A COUPLE OF YEARS LATER, ASCANIO SOBRERO (ITALIAN, 1812-88) DISCOVERED *NITROGLYCERINE!*

ORIGINALLY I CALLED IT *PYROGLYCERINE*!*

*FROM GREEK πῦρ ("PÛR", FIRE)

NITROGLYCERINE WAS USED TO BLAST ROCKS FOR CONSTRUCTION THROUGH MOUNTAINS!

YET NITROGLYCERINE WAS *SENSITIVE!*

THE FAMILY OF ALFRED BERNHARD NOBEL (SWEDISH, 1833–96) RAN A NITROGLYCERINE FACTORY...

Business is Booming!

THEY SENT ALFRED TO PARIS IN 1850, WHERE HE MET SOBRERO...

IN 1864, ALFRED'S YOUNGER BROTHER EMIL WAS KILLED IN A FACTORY EXPLOSION!

KAPOW!

ALFRED GOT UPSET...

HOW CAN I DESENSITIZE NITROGLYCERIN?

BY 1867, NOBEL FOUND AN ANSWER—

DYNAMITE!

FUSE
GUNPOWDER IN IGNITER
ABSORBENT MATERIAL (DIATOMACEOUS EARTH OR SAWDUST) SOAKED IN NITROGLYCERIN
CARDBOARD CASE

MUCH SAFER AND LESS SHOCK-SENSITIVE, DYNAMITE WAS ALSO USED FOR BLASTING AND CONSTRUCTION...

... PLUS COUNTLESS 20TH-CENTURY TELEVISION CARTOONS!

DETONATOR

THE GREAT WAR* BEGAN IN 1914...

*LATER CALLED "WORLD WAR ONE"

BY SPRING 1915, GERMANY BEGAN TO USE POISON GASES AGAINST ALLIED TROOPS!

Helft uns fiegen!

THE FIRST INCIDENT WAS APRIL 22, NEAR YPRES, BELGIUM...

HOLLAND
YPRES
BELGIUM
GERMANY
FRANCE

THEY RELEASED TONS OF CHLORINE GAS OVER A 4-MILE STRETCH...

Cl_2 GAS, WITH A GREENISH CLOUD AND SHARP ODOR...

INJURED 7,000...
KILLED 1,100...

...VIOLATING INTERNATIONAL LAW!

THE BRITISH RESPONDED LIKEWISE AT THE BATTLE OF LOOS, FRANCE ON SEPTEMBER 25, BUT THE CHLORINE DIDN'T DISPERSE WELL...

UK
BELGIUM
FRANCE
LOOS

PHOSGENE, $COCl_2$, WAS STRONGER: COLORLESS, SMELLED LIKE MOLDY HAY...

...THE GERMANS DEPLOYED PHOSGENE ON DECEMBER 19 AGAINST THE BRITISH!

MORE INFAMOUS WAS *MUSTARD GAS*, WHICH DISPERSED AS AN OILY LIQUID, BLISTERING SKIN, EYES, AND BRONCHIAL TUBES!

$$S \Big\langle \begin{array}{l} CH_2CH_2Cl \\ CH_2CH_2Cl \end{array}$$

THE GERMANS USED IT IN JULY 1917!

BY 1918, THE ALLIES RE-SPONDED WITH MUSTARD GAS ATTACKS ON THE GERMANS!

REMEMBER BELGIUM
Buy Bonds
Fourth
Liberty
Loan

TOTAL CASUALTIES IN WORLD WAR I FROM GAS WARFARE? NEARLY 100,000...

...MOSTLY FROM PHOSGENE!

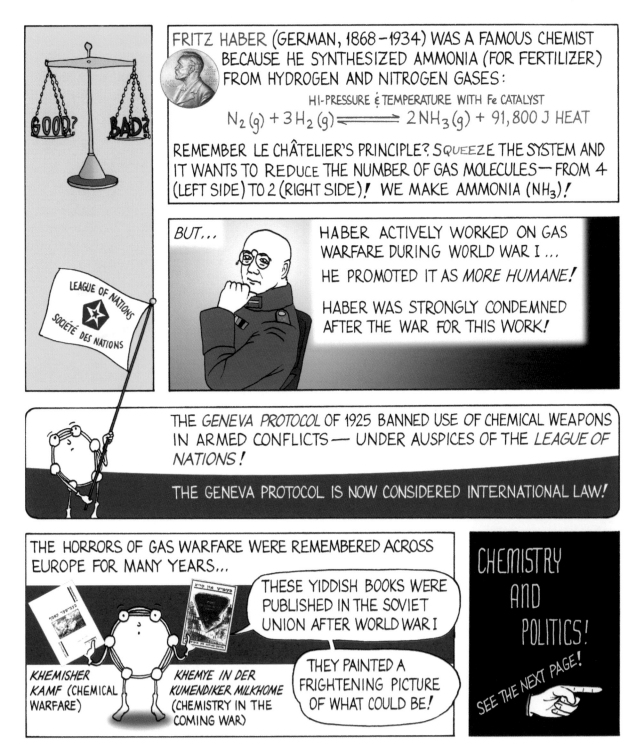

IN 1936, GERHARD SCHRADER (GERMAN, 1903-90) WAS A CHEMIST WORKING FOR THE FIRM I.G. FARBEN, FINDING INSECTICIDES TO PRESERVE THE FOOD SUPPLY— AT THE BEHEST OF THE NAZIS!

LET'S TRY A PHOSPHORUS ATOM PLUS A CYANIDE GROUP!

THE RESULT ?

AN ORGANOPHOSPHORUS COMPOUND THAT KILLED INSECTS *AND PEOPLE!*

SCHRADER CALLED IT "PREPARATION 9/91"

I.G. FARBEN NOTIFIED THE NAZI MILITARY...

WELL, WELL, WELL... TAKE A LOOK AT *THIS!*

THE NAZIS RENAMED IT *TABUN*, (FROM "TABOO"), BECAUSE IT IS INCREDIBLY TOXIC:

ACETYLCHOLINE →
NERVE
MUSCLE
CONTRACTS

ACETYLCHOLINE IS THE MESSENGER MOLECULE FROM BRAIN TO MUSCLE; AFTER RECEIPT, *ACETYLCHOLINES-TERASE* BREAKS IT DOWN...

...BUT TABUN BLOCKS ACETYLCHOLINESTERASE!

THE MUSCLE *NEVER STOPS* CONTRACTING!

THUS TABUN IS A *"NERVE AGENT"!*

TWO YEARS LATER SCHRADER CREATED ANOTHER, DEADLIER NERVE AGENT:

THE NAZIS NAMED IT *SARIN...*

...AFTER THE CREATORS: SCHRADER, OTTO AMBROSE, GERHARD RITTER, AND HANS-JÜRGEN VON DER LINDE.

TABUN AND SARIN ARE MEMBERS OF THE *G-SERIES* (FOR "GERMAN") OF NERVE AGENTS ...

ALSO USED IN
IRAQ   1988
JAPAN   1995
SYRIA   2017

THE BRITISH INVENTED THE "V-SERIES" OF NERVE AGENTS IN THE 1950s...

(V FOR "VENOMOUS")

...AND THE RUSSIANS INVENTED THE *NOVICHOK** SERIES IN THE 1960s – 90s!

*NOVICHOK (НОВИЧÓК) MEANS "NEWCOMER"

CHAPTER 12:
ELECTRONS AND PROTONS AND NEUTRONS, OH MY!

19TH-CENTURY SCIENTISTS FOUND THAT THEY COULD FORCE ELECTRICITY TO MOVE THROUGH NEARLY ANY MATERIAL...

AS LONG AS YOU PUSH HARD ENOUGH!

BUT DOES ELECTRICITY GO THROUGH A VACUUM?

FIRST YOU NEED TO MAKE A RELIABLE VACUUM!

UM, NO.

JOHANN HEINRICH WILHELM GEISSLER (GERMAN, 1814-79) INVENTED A PUMP THAT CREATED A BETTER VACUUM IN 1857...

...AND GLASS TUBES THAT COULD RETAIN A VACUUM!

HIS COLLEAGUE, JULIUS PLÜCKER (GERMAN, 1801-68), USED THESE GEISSLER TUBES TO EXAMINE THE GLOW WHEN TWO ELECTRODES WERE PLACED INSIDE:

HIGH VOLTAGE

PLÜCKER FOUND THAT A MAGNET DEFLECTS THE GLOW!

COME ON, HERR PLÜCKER, SMILE FOR THE DAGUERROTYPE!

CHAPTER 13: ATOMIC STRUCTURE & MOLECULAR BONDING

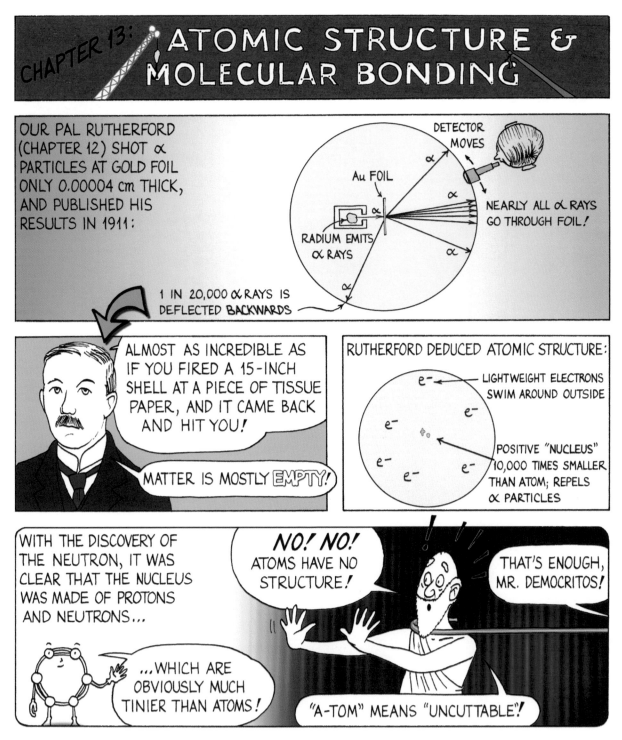

OUR PAL RUTHERFORD (CHAPTER 12) SHOT α PARTICLES AT GOLD FOIL ONLY 0.00004 cm THICK, AND PUBLISHED HIS RESULTS IN 1911:

DETECTOR MOVES

Au FOIL

RADIUM EMITS α RAYS

NEARLY ALL α RAYS GO THROUGH FOIL!

1 IN 20,000 α RAYS IS DEFLECTED BACKWARDS

ALMOST AS INCREDIBLE AS IF YOU FIRED A 15-INCH SHELL AT A PIECE OF TISSUE PAPER, AND IT CAME BACK AND HIT YOU!

MATTER IS MOSTLY EMPTY!

RUTHERFORD DEDUCED ATOMIC STRUCTURE:

LIGHTWEIGHT ELECTRONS SWIM AROUND OUTSIDE

POSITIVE "NUCLEUS" 10,000 TIMES SMALLER THAN ATOM; REPELS α PARTICLES

WITH THE DISCOVERY OF THE NEUTRON, IT WAS CLEAR THAT THE NUCLEUS WAS MADE OF PROTONS AND NEUTRONS...

...WHICH ARE OBVIOUSLY MUCH TINIER THAN ATOMS!

NO! NO! ATOMS HAVE NO STRUCTURE!

THAT'S ENOUGH, MR. DEMOCRITOS!

"A-TOM" MEANS "UNCUTTABLE"!

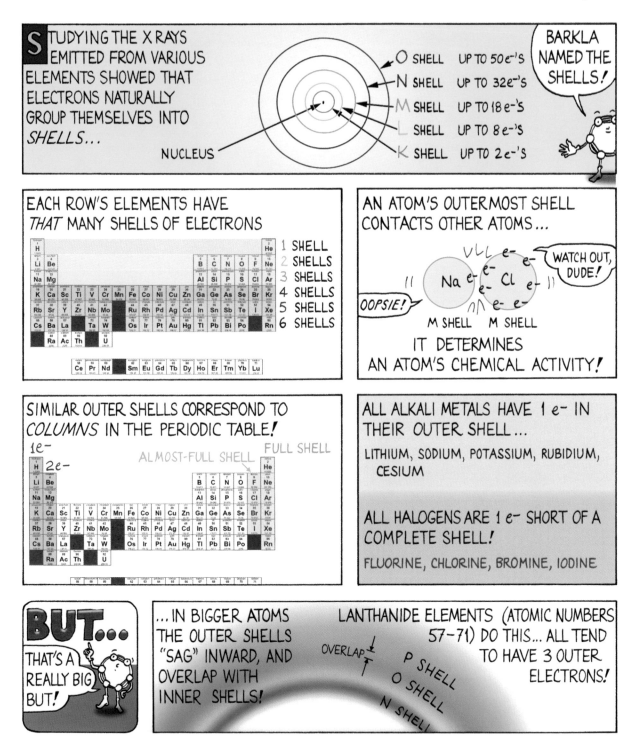

STUDYING THE X RAYS EMITTED FROM VARIOUS ELEMENTS SHOWED THAT ELECTRONS NATURALLY GROUP THEMSELVES INTO *SHELLS*...

NUCLEUS

O SHELL UP TO 50 e^-'s
N SHELL UP TO 32 e^-'s
M SHELL UP TO 18 e^-'s
L SHELL UP TO 8 e^-'s
K SHELL UP TO 2 e^-'s

BARKLA NAMED THE SHELLS!

EACH ROW'S ELEMENTS HAVE *THAT* MANY SHELLS OF ELECTRONS

1 SHELL
2 SHELLS
3 SHELLS
4 SHELLS
5 SHELLS
6 SHELLS

AN ATOM'S OUTERMOST SHELL CONTACTS OTHER ATOMS...

WATCH OUT, DUDE!

OOPSIE!

M SHELL   M SHELL

IT DETERMINES AN ATOM'S CHEMICAL ACTIVITY!

SIMILAR OUTER SHELLS CORRESPOND TO *COLUMNS* IN THE PERIODIC TABLE!

1e^-
2e^-
ALMOST-FULL SHELL
FULL SHELL

ALL ALKALI METALS HAVE 1 e^- IN THEIR OUTER SHELL...

LITHIUM, SODIUM, POTASSIUM, RUBIDIUM, CESIUM

ALL HALOGENS ARE 1 e^- SHORT OF A COMPLETE SHELL!

FLUORINE, CHLORINE, BROMINE, IODINE

BUT... THAT'S A REALLY BIG BUT!

...IN BIGGER ATOMS THE OUTER SHELLS "SAG" INWARD, AND OVERLAP WITH INNER SHELLS!

OVERLAP
P SHELL
O SHELL
N SHELL

LANTHANIDE ELEMENTS (ATOMIC NUMBERS 57–71) DO THIS... ALL TEND TO HAVE 3 OUTER ELECTRONS!

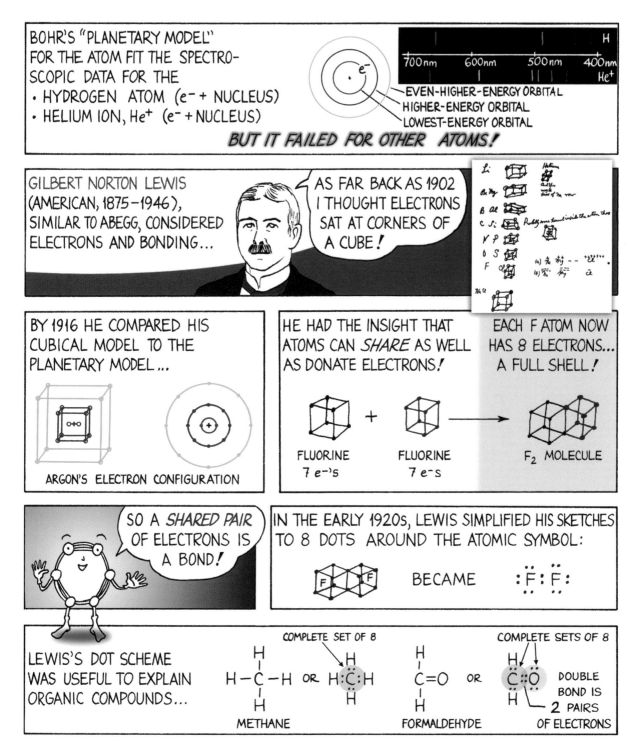

BOHR'S "PLANETARY MODEL" FOR THE ATOM FIT THE SPECTRO-SCOPIC DATA FOR THE
- HYDROGEN ATOM (e^- + NUCLEUS)
- HELIUM ION, He^+ (e^- + NUCLEUS)

EVEN-HIGHER-ENERGY ORBITAL
HIGHER-ENERGY ORBITAL
LOWEST-ENERGY ORBITAL

BUT IT FAILED FOR OTHER ATOMS!

GILBERT NORTON LEWIS (AMERICAN, 1875–1946), SIMILAR TO ABEGG, CONSIDERED ELECTRONS AND BONDING...

AS FAR BACK AS 1902 I THOUGHT ELECTRONS SAT AT CORNERS OF A CUBE!

BY 1916 HE COMPARED HIS CUBICAL MODEL TO THE PLANETARY MODEL...

ARGON'S ELECTRON CONFIGURATION

HE HAD THE INSIGHT THAT ATOMS CAN *SHARE* AS WELL AS DONATE ELECTRONS!

FLUORINE 7 e^-'s

FLUORINE 7 e^-s

EACH F ATOM NOW HAS 8 ELECTRONS... A FULL SHELL!

F_2 MOLECULE

SO A *SHARED PAIR* OF ELECTRONS IS A BOND!

IN THE EARLY 1920s, LEWIS SIMPLIFIED HIS SKETCHES TO 8 DOTS AROUND THE ATOMIC SYMBOL:

BECAME

LEWIS'S DOT SCHEME WAS USEFUL TO EXPLAIN ORGANIC COMPOUNDS...

COMPLETE SET OF 8

METHANE

FORMALDEHYDE

COMPLETE SETS OF 8

DOUBLE BOND IS 2 PAIRS OF ELECTRONS

LEWIS'S DOT STRUCTURES EVEN HELP US UNDERSTAND *TRIPLE BONDS!*

$H-C\equiv C-H$ OR

ACETYLENE

2 COMPLETE SETS OF 8 ELECTRONS

H:C:::C:H

3 SHARED PAIRS OF ELECTRONS

IRVING LANGMUIR (AMERICAN, 1881–1957) EXTENDED LEWIS'S IDEAS...

WE CALL HIS 8-DOT SYSTEM THE *OCTET RULE*...

...AND BONDS WHERE ELECTRONS ARE SHARED, *COVALENT BONDS!*

LEWIS CONTINUED WITH HIS NOTION OF ELECTRONS, TO FORM A NEW THEORY OF ACIDS AND BASES, IN 1923...

ACIDS *ACCEPT* ELECTRON-PAIRS, BASES *DONATE* ELECTRON-PAIRS!

BASE :   ACID
H_3N + BF_3 $\longrightarrow$ NH_3BF_3
AMMONIA   BORON TRIFLUORIDE

PERSONAL POLITICS

OR

YOU NEVER KNOW WHO MIGHT GET UPSET...

LEWIS SPENT TIME IN 1901 IN NERNST'S LAB...

THEY DIDN'T GET ALONG!

NERNST'S WORK ON HEAT IS A REGRETTABLE EPISODE IN THE HISTORY OF CHEMISTRY

FRECHER KERL! I'LL TELL MY FRIENDS ON THE NOBEL PRIZE COMMITTEE THAT LEWIS SHALL *NOT* GET THE PRIZE!

IN 1919, MAURICE LOYAL HUGGINS (AMERICAN, 1897-1981) WAS AN UNDERGRAD, AND STUDENT OF LEWIS. HE WAS TRYING TO UNDERSTAND THIS REACTION...

KETO FORM

INTERCONVERSION

ENOL FORM

MAYBE A WEIRD BOND EXISTS IN THE ENOL FORM...?

WEIRD 3-WAY BOND

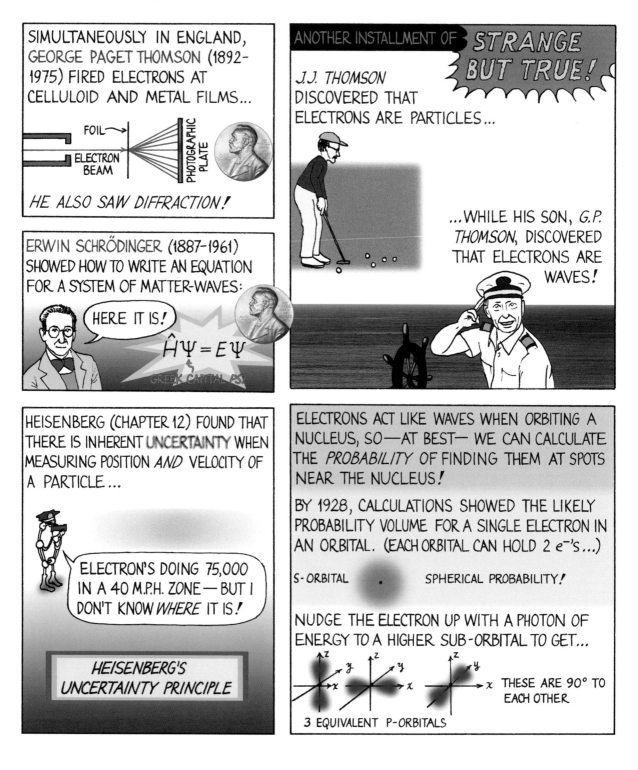

LINUS PAULING (AMERICAN, 1901–94) DECIDED TO BRING THE IDEAS OF QUANTUM MECHANICS TO CHEMISTRY WITH A SERIES OF 7 PAPERS, STARTING IN 1931...

LOOK AT A CARBON ATOM, WITH 4 VALENCE ELECTRONS...

1 e^- IS IN THE 2S ORBITAL

3 e^-'S ARE IN THE 2P ORBITALS

CONSIDER THESE 4 ELECTRONS AS *EQUIVALENT!*

THE 4 ORBITALS MIX TOGETHER, OR *HYBRIDIZE*, TO MAKE NEW SHAPES...

3 p ORBITALS   S ORBITAL

"sp^3" ORBITALS

THE SHAPE OF THE 4 sp^3 ORBITALS IS *TETRAHEDRAL*

JUST AS VAN'T HOFF AND LE BEL SUGGESTED!

METHANE, CH_4, THEN LOOKS LIKE THIS:

OVERLAPS ARE BONDS

WITH ETHYLENE, $H_2C = CH_2$, REHYBRIDIZE THE ORBITALS:

ONLY 3 OF THE 4 VALENCE ELECTRONS MIX...

P   P   S

"sp^2" ORBITALS

LEFTOVER ELECTRONS MAKE A SECOND BOND

FOR BENZENE, RESONANCE IS EXPLAINED:

NO FLIPPING BACK & FORTH!

INSTEAD, BENZENE IS A HYBRID OF *BOTH* STRUCTURES:

THE HEXAGONAL RING IS FORMED WITH sp^2 ORBITALS AS IN ETHYLENE, *BUT...*

THE LEFTOVER P-ORBITALS ALL INTERACT!

WHO KNEW I AM SO COMPLICATED?

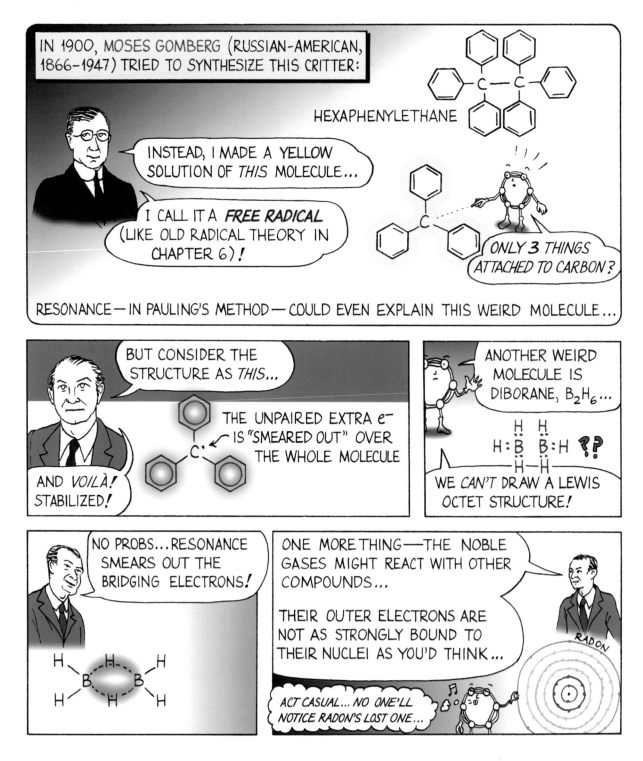

NEIL BARTLETT (BRITISH-AMERICAN, 1932-2008)

IN 1962 I MANAGED TO SYNTHESIZE THE FIRST NOBLE GAS COMPOUND, $Xe^+ [PtF_6]^-$

(IN 2000, BARTLETT SUGGESTED IT MIGHT REALLY BE A *MIXTURE* OF XENON SALTS)

ONCE CONSIDERED INERT, THE RIGHTMOST COLUMN OF ELEMENTS, FROM HELIUM TO RADON, ARE NOW KNOWN TO FORM IONS AND COMPOUNDS!

XeF_2
KrF_4
$HArF$
RnF_2
ArH^+
KrF_2
RnO_3
$HeNe^+$
HeH^+
$NeAr^+$

PAULING WAS RIGHT!

PAULING WON HIS 2ND NOBEL PRIZE—THE PEACE PRIZE—FOR ANTI-NUCLEAR-WAR EFFORTS!

WASHINGTON, DC, APRIL 28, 1962 →

Mr. Kennedy
Mr. Macmillan
WE HAVE NO RIGHT TO TEST

IN THE LAST DECADES OF HIS LIFE, PAULING WIDELY PROMOTED THE IDEA THAT LARGE DOSES OF VITAMIN C PREVENT COLDS AND CANCER...

SADLY, THE EVIDENCE DOES NOT SUPPORT IT!

MOLECULAR ORBITAL THEORY DEVELOPED SIMULTANEOUSLY WITH PAULING'S METHOD, LARGELY THROUGH 4 PEOPLE:

THE FAB FOUR!

FRIEDRICH HUND (GERMANY, 1896-1997)
ROBERT MULLIKAN (USA, 1896-1986)
JOHN SLATER (USA, 1900-76)
JOHN LENNARD-JONES (UK, 1894-1954)

MOLECULAR ORBITAL THEORY ATTEMPTS TO CALCULATE THE ACTUAL SHAPES OF THE PROBABILITY VOLUMES FOR ORBITALS WHEN YOU BRING ATOMS TOGETHER...

SO FOR 2 HYDROGEN ATOMS
1s   MAKES   1s
H_2
OVERALL NEW MOLECULAR ORBITAL:
→ σ BONDING ORBITAL
GREEK LOWER-CASE SIGMA

ELECTRONS ARE WAVES, SO CAN BE OUT OF PHASE
1s   MAKES   1s
H_2
σ^* ANTI-BONDING ORBITAL

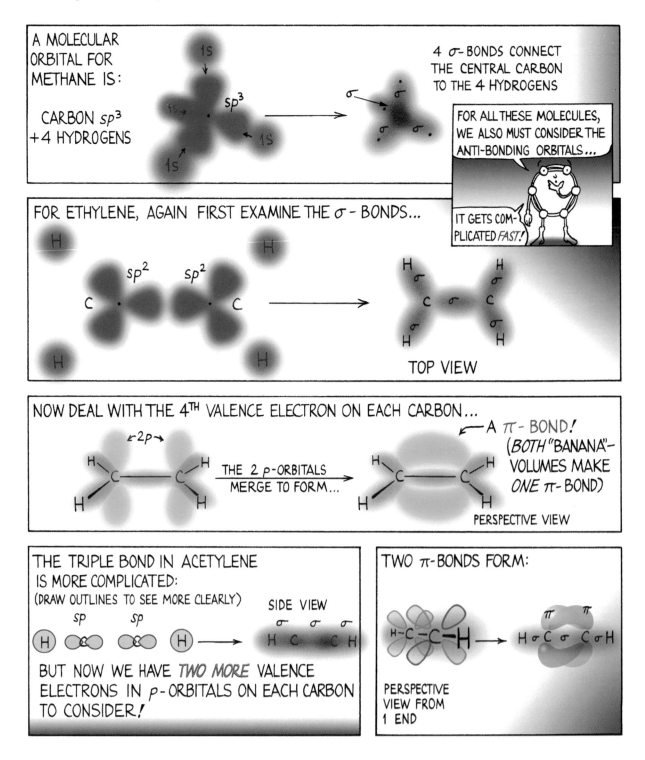

A MOLECULAR ORBITAL FOR METHANE IS:

CARBON sp^3 + 4 HYDROGENS

4 σ-BONDS CONNECT THE CENTRAL CARBON TO THE 4 HYDROGENS

FOR ALL THESE MOLECULES, WE ALSO MUST CONSIDER THE ANTI-BONDING ORBITALS...

IT GETS COMPLICATED *FAST!*

FOR ETHYLENE, AGAIN FIRST EXAMINE THE σ-BONDS...

TOP VIEW

NOW DEAL WITH THE 4TH VALENCE ELECTRON ON EACH CARBON...

THE 2 p-ORBITALS MERGE TO FORM...

A π-BOND! (*BOTH* "BANANA"-VOLUMES MAKE *ONE* π-BOND)

PERSPECTIVE VIEW

THE TRIPLE BOND IN ACETYLENE IS MORE COMPLICATED: (DRAW OUTLINES TO SEE MORE CLEARLY)

SIDE VIEW

BUT NOW WE HAVE *TWO MORE* VALENCE ELECTRONS IN p-ORBITALS ON EACH CARBON TO CONSIDER!

TWO π-BONDS FORM:

PERSPECTIVE VIEW FROM 1 END

SO HOW DOES MOLECULAR ORBITAL THEORY TREAT *ME?*

TAP TAP

WE START WITH THE σ-BONDS FOR THE BASIC HEXAGON:

EACH CARBON IS sp^2 HYBRIDIZED

AND EACH CARBON HAS ONE LEFTOVER ELECTRON IN A p-ORBITAL...

THE p-ORBITALS ALL MERGE TO BECOME A π-ORBITAL!

NOT ONLY DO I *LOVE* PIE, BUT I *AM* PI!

PAULING'S METHOD AND MOLECULAR ORBITAL THEORY USUALLY AGREE... BUT SOMETIMES MOLECULAR ORBITAL THEORY WORKS BETTER:

CASE IN POINT   OXYGEN, O_2

THE LEWIS STRUCTURE IS :Ö::Ö: WITH ALL ELECTRONS PAIRED

PAULING'S SCHEME ALSO PREDICTS ALL ELECTRONS PAIRED:

BUT *IN REALITY*, OXYGEN IS *PARAMAGNETIC* —THE MOLECULE IS WEAKLY ATTRACTED TO A MAGNET!

LIQUID O_2 STAYS IN THE MAGNETIC FIELD!

N   S

O_2 MOLECULES HAVE *UNPAIRED ELECTRONS!* (AND LIQUID O_2 IS PALE BLUE!)

MOLECULAR ORBITAL THEORY SHOWS THIS!

THE TWO UNPAIRED ELECTRONS ARE IN AN ANTI-BONDING ORBITAL!

MOLECULAR ORBITAL THEORY ONLY BECAME POPULAR WITH POWERFUL COMPUTERS!

CHAPTER 14: POLYMERS

ENGLAND IN THE 19TH CENTURY: A PERFECT PLACE TO WEAR A MACKINTOSH!

(RUBBERIZED RAINCOAT!)

RUBBERIZED APPAREL CAME TO AMERICA IN THE 1830s...

BLEAH!

STRETCH...

BUT THE RUBBER MELTED IN THE SUMMER AND HARDENED IN THE WINTER!

AMERICANS NATHANIEL HAYWARD (1808-65) AND CHARLES GOODYEAR (1800-60) ACCIDENTALLY DROPPED SULFUR-COATED RUBBER ON A HOT STOVE.... BY 1844 THEY PATENTED A BETTER *VULCANIZED RUBBER!*

CHECK THIS OUT, DUDE!

S-T-R-E-T-C-H

GOODYEAR

HAYWARD

A DECADE LATER, ALEXANDER PARKES (ENGLISH, 1813–1900) INVENTED A MOLDABLE SOLID MADE FROM NITROCELLULOSE (SEE CHAPTER 11) AND WOOD NAPHTHA (METHANOL)...

THE FIRST PLASTIC!

AT THE LONDON INTERNATIONAL EXHIBITION, 1862...

I CALL IT *PARKESINE!*

ÉLEUTHÈRE IRÉNÉE DU PONT DE NEMOURS (FRENCH-AMERICAN, 1771–1834) WAS A STUDENT OF LAVOISIER (CHAPTER 4) IN MANUFACTURING GUNPOWDER!

HE FLED FRANCE TO AMERICA AFTER THE FRENCH REVOLUTION...

BIENVENU AUX ÉTATS-UNIS!

RHODE ISLAND, 1800

...THEN FOUNDED A GUNPOWDER FACTORY IN GREENVILLE, DELAWARE!

VISIT IT AT THE HAGLEY MUSEUM!

CAROTHERS'S FIRST SAMPLE OF A POLYMER FIBER WAS DRAWN FROM A SOLUTION OF ADIPIC ACID AND HEXAMETHYLENEDIAMINE IN FEBRUARY 1935...

"FIBER 66"

6 CARBONS

6 CARBONS

LARGE NUMBER

ADIPIC ACID   HEXAMETHYLENEDIAMINE

DUPONT COMPANY TOSSED AROUND IDEAS TO NAME THIS POLYMER...

DUPAROOH? (DUPONT PULLS A RABBIT OUT OF A HAT)
WACARA? (WALLACE CAROTHERS)
DELAWEAR?

NO.
NO!
HECK NO!

HOW ABOUT NURON ("NO-RUN," BUT BACKWARDS)?

TOO MUCH LIKE "NEURON"!

NYLON?

UM... OKAY!

DUPONT PATENTED NYLON IN 1938...

NYLON STOCKINGS

A SENSATION AT THE 1939 WORLD'S FAIR!

OTHER USES FOR NYLON:

CARPETS

GUITAR STRINGS

ON TOP OF OL' SMOKEY, ALL COVERED WITH CHEESE

MACHINE PARTS

PLUNK

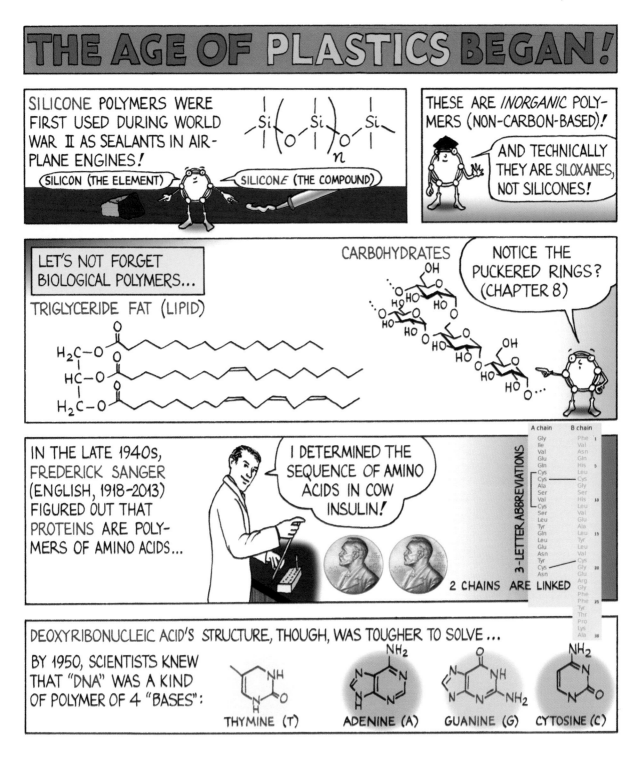

ROSALIND ELSIE FRANKLIN (ENGLISH, 1920–58) WAS AN EXPERT X-RAY CRYSTALLOGRAPHER SPECIALIZING IN IMAGES OF IMPERFECT CRYSTALS...

IN 1952 SHE TOOK THIS X-RAY DIFFRACTION IMAGE OF DNA... BASED ON IT SHE GUESSED THAT DNA WAS LIKELY A HELIX!

IN JANUARY 1953, HER COLLEAGUE, MAURICE WILKINS, SHOWED JAMES DEWEY WATSON (AMERICAN, b. 1928) THE IMAGE—*WITHOUT HER KNOWING!*

WATSON AND FRANCIS HARRY COMPTON CRICK (ENGLISH, 1916–2004) RUSHED TO BUILD A MODEL OF DNA!

CRICK

WATSON

A DNA FIBER IS TWO STRANDS OF DNA POLYMER, ATTACHED TOGETHER WITH HYDROGEN BONDS (CHAPTER 13) BETWEEN PAIRS OF BASES:

GUANINE (G)   CYTOSINE (C)   ADENINE (A)   THYMINE (T)

POLYETHYLENE WAS DIFFICULT TO MAKE...

HIGH TEMPERATURES

CRUNCH

HIGH PRESSURES

KARL WALDEMAR ZIEGLER (GERMAN, 1898–1973) AND GIULIO NATTA (ITALIAN, 1903–79) DEVELOPED SALTS OF TITANIUM, CHROMIUM, AND ZIRCONIUM THAT CATALYZED THE REACTION!

ZIEGLER   NATTA

1950   NOW

PRODUCTION

PRODUCTION INCREASED!

Chapter 15

ISOTOPES AND ARTIFICIAL ELEMENTS

LET'S LOOK IN ON WILLIAM CROOKES (CHAPTER 12) IN 1900...

THIS IS BIZARRE... *FRESH* URANIUM COMPOUNDS ARE ONLY WEAKLY RADIOACTIVE ...

BUT THOSE LEFT FOR A WHILE ARE *MORE* RADIOACTIVE!

FREDERICK SODDY (ENGLISH, 1877–1956) POINTED OUT IN 1902...

1) URANIUM ATOMS EMIT α- PARTICLES...

2) THE URANIUM ATOMS THUS *TRANSMUTE* INTO OTHER ELEMENTS WITH *STRONGER* RADIOACTIVITY...

... AND 3) THESE OTHER ELEMENTS RADIOACTIVELY DECAY, SUCH AS THIS KNOWN SERIES:

ACTINIUM $\xrightarrow{\alpha}$ RADIOACTINIUM $\xrightarrow{}$ ACTINIUM X $\xrightarrow{\alpha}$ ACTINIUM EMANATION $\xrightarrow{\alpha}$ ACTINIUM A $\xrightarrow{\alpha}$

ALL THESE SERIES END WITH LEAD!

LEAD $\xleftarrow{\alpha}$ ACTINIUM D $\xleftarrow{}$ ACTINIUM C $\xleftarrow{\alpha}$ ACTINIUM B

...WHICH VIOLATES ROBERT BOYLE'S IDEA OF ELEMENTS!

YOU WIN SOME, AND YOU LOSE SOME...

IN 1907, RUTHERFORD (CHAPTER 12) DISCOVERED THAT EACH ELEMENT HAS ITS OWN RATE OF RADIOACTIVE DECAY:

$t_{1/2}$, HALF-LIFE...
TIME FOR ELEMENT TO HALVE ITS AMOUNT

$y = e^{-ax}$

AMOUNT

$t_{1/2}$

ANOTHER $t_{1/2}$

TIME

THORIUM'S HALF-LIFE = 14 BILLION* YEARS
URANIUM'S HALF-LIFE = 4.5 BILLION YEARS
RADIUM'S HALF-LIFE = 1,600 YEARS

*AMERICAN BILLION, 10^9, 1,000,000,000

RUTHERFORD THEN SLAMMED DEUTERIUM NUCLEI TOGETHER IN 1934...

$$^2H + {}^2H \longrightarrow {}^3H + {}^1H$$

...AND MADE EVEN *HEAVIER* HYDROGEN, OR ***TRITIUM*** *

...WHICH IS RADIOACTIVE*!*

$$^3H \xrightarrow{e^-} {}^3He \qquad t_{1/2} = 12.3 \text{ YEARS}$$

*GREEK τρίτος (*TRITOS*, "THIRD")

NEUTRONS IN THE NUCLEUS CAUSE THE DIFFERENCE IN MASS AMONG ISOTOPES*!*

PROTIUM (REGULAR HYDROGEN) ATOMIC WEIGHT = 1

DEUTERIUM ATOMIC WEIGHT = 2

TRITIUM ATOMIC WEIGHT = 3

IN 1925, PATRICK MAYNARD BLACKETT (ENGLISH, 1897–1974) PHOTOGRAPHED ION TRACKS IN A CLOUD CHAMBER...

THIS PHOTO SHOWS ARTIFICIAL TRANSMUTATION OF NITROGEN INTO OXYGEN*!*

$$^{14}_7N + {}^4_2He \longrightarrow {}^{17}_8O + {}^1_1H$$

α - PARTICLE   HEAVY OXYGEN

THE ANCIENT DREAM OF **ALCHEMY** *CAME TRUE!*

BUT NATURAL α-PARTICLES ARE RELATIVELY WEAK... CAN WE JACK UP THEIR ENERGY?

JOHN DOUGLAS COCKCROFT (ENGLISH, 1897–1967) AND ERNEST THOMAS SINTON WALTON (IRISH, 1903–95) BUILT A *PARTICLE ACCELERATOR* IN 1932 TO GIVE PROTONS HIGHER ENERGIES...

$$^7_3Li + {}^1_1H \longrightarrow {}^4_2He + {}^4_2He$$

HIGH-ENERGY PROTON

MODERN PARTICLE ACCELERATORS CONTINUE THIS IDEA*!*

TO NUCLEAR PHYSICS

LET'S MAKE NEW RADIOACTIVE ATOMS*!*

FRÉDÉRIC   IRÈNE
JOLIOT–CURIE
(1900–58)   (1897–1956)

(PHOSPHORUS–31 IS THE REGULAR ISOTOPE)

$$^{27}Al + {}^2He \longrightarrow {}^{30}P + n$$

RADIOACTIVE PHOSPHORUS; $t_{1/2} = 2.5$ MINUTES

IT'S 1935, AND WE HAVE 3 GAPS IN THE PERIODIC TABLE...

...CAN WE MAKE ELEMENTS 43, 85, AND 87?

IN 1937, ERNEST ORLANDO LAWRENCE (AMERICAN, 1901–58) BOMBARDED MOLYBDENUM FOIL WITH DEUTERIUM IONS...

I MAILED IT TO MY COLLEAGUE IN ITALY...

EMILIO GINO SEGRÈ (1905-89) FOUND ELEMENT 43!

CIAO! I CALL IT *TECHNETIUM**! WE FOUND ISOTOPES 95, 97, AND —LATER—99!

^{99}Tc IS NOW USED AS A RADIOACTIVE TRACER IN MEDICINE!

*GREEK τεχνητός (*TEKHNITOS*, "ARTIFICIAL")

IN 1939, MARGUERITE CATHERINE PEREY (FRANCE, 1909–75) DISCOVERED ELEMENT 87 THROUGH α-DECAY OF ACTINIUM...

$$^{227}Ac \xrightarrow{\alpha} {}^{223}Fr$$

I NAME IT *CATIUM**... EUH...NON...*FRANCIUM!*

*REALLY!

THE NEXT YEAR SEGRÈ SYNTHESIZED ELEMENT 85 USING BISMUTH...

$$^{209}_{83}Bi + {}^{4}_{2}He \rightarrow {}^{211}_{85}At + 2n$$

HE NAMED IT *ASTATINE**

*GREEK ἄστατος (*ASTATOS*, "UNSTABLE")

AND THE **PERIODIC TABLE** WAS *COMPLETE!*

WELL...

CAN WE MAKE ELEMENTS *AFTER URANIUM* IN THE PERIODIC TABLE?

IN 1940, AMERICANS EDWIN MATTISON MC-MILLAN (1907-91) AND PHILIP HAUGE ABELSON (1913-2004) DID THAT!

$$^{238}_{92}U + n \longrightarrow {}^{239}_{92}U \xrightarrow[t_{1/2} = 23 \text{ MIN}]{e-} {}^{239}_{93}Np$$

WE CALL IT *NEPTUNIUM**

*IT COMES AFTER URANIUM, AS NEPTUNE FOLLOWS URANUS

THROUGH THE 1940s AND '50s, GLENN THEODORE SEABORG (AMERICAN, 1912-99) FOUND MORE *TRANSURANIUM ELEMENTS...*

94	PLUTONIUM — *ATOMIC POWER*	98	CALIFORNIUM
95	AMERICIUM — *SMOKE DETECTORS*	99	EINSTEINIUM
96	CURIUM — *ELECTRICITY IN SPACECRAFT*	100	FERMIUM
97	BERKELIUM	101	MENDELEVIUM

STUDYING THEIR CHEMISTRY, SEABORG REALIZED THAT...

ELEMENTS 89 AND UP ARE CHEMICALLY SIMILAR LIKE THE LANTHANIDES ARE...

...SO LET'S GIVE THEM THEIR OWN ROW!

AND THE MODERN FORM OF THE PERIODIC TABLE WAS CREATED!

SEABORG WAS THE 1ST *LIVING* PERSON WITH AN ELEMENT NAMED FOR HIM!

SEABORGIUM (ELEMENT 106)

106 Sg

THE SECOND LIVING PERSON FOR WHOM AN ELEMENT IS NAMED IS YURI TSOLAKOVICH OGANESSIAN (RUSSIAN-ARMENIAN, b. 1933):

ELEMENTS OGANESSIAN DISCOVERED (OR HELPED DISCOVER)

106	SEABORGIUM	114	FLEROVIUM
107	BOHRIUM	115	MOSCOVIUM
108	HASSIUM	116	LIVERMORIUM
110	DARMSTADTIUM	117	TENNESSINE

118 *OGANESSON*

HALF-LIVES FOR THESE *SUPERHEAVY ELEMENTS* GET SHORTER AS YOU INCREASE ATOMIC NUMBER!

IT GETS HARDER TO TRAP THESE!

THIS IS NOT A LINEAR, BUT A LOGARITHMIC SCALE

HALF-LIFE (SECONDS)

10,000
100
1
0.01

105   107   109   111   113   115   117
ATOMIC NUMBER

CHAPTER 16
ENVIRONMENTAL CHEMISTRY

BY 1800, SCIENTISTS KNEW THAT PLANTS TAKE IN CO_2 AND GIVE OFF O_2, WHILE ANIMALS DO THE OPPOSITE...

NICOLAS-THÉODORE DE SAUSSURE (SWISS, 1767-1845) DISCOVERED THAT THE CARBON FROM CO_2 IS INCORPORATED INTO PLANTS.

AIR WITH 10% CO_2

ADOLPHE-THÉODORE BRONGNIART (FRENCH, 1801-76) NOTED THAT COAL (CARBON) IS BIOLOGICAL IN ORIGIN, SO CARBON GETS FIXED INTO THE EARTH!

OH LÀ LÀ! CHECK OUT THIS FOSSIL!

JEAN-BAPTISTE BOUSSINGAULT (FRENCH, 1801-87) SAMPLED VOLCANIC GASES AND FOUND *LOTS* OF CO_2!

SOME CO_2 CLEARLY COMES FROM GEOLOGICAL SOURCES!

JACQUES-JOSEPH ÉBELMEN (FRENCH, 1814-52) PROPOSED IN THE LATE 1840s THAT CO_2 IN SOIL AND CARBONATE ROCKS UNDERGOES CHEMICAL WEATHERING...

THE REACTION IS $CaSiO_3 + CO_2 \rightarrow CaCO_3 + SiO_2$
CALCIUM METASILICATE   CALCIUM CARBONATE   SILICON DIOXIDE

THERE IS A CONTINUAL *CARBON ROTATION* ON THE EARTH!

ALSO IN THE 1920S REFRIGERATORS ENTERED HOME KITCHENS!

EXCEPT THE RE-FRIGERANT WAS SOMETHING TOXIC OR FLAMMABLE LIKE NH_3, SO_2, OR PROPANE!

IN 1930, MIDGELY UNVEILED FREON® ($DICHLORODIFLUORO$METHANE, CCl_2F_2)...

IT'S NONFLAMMABLE & NON-TOXIC!

THE REFRIGERATION AND AIR-CONDITIONING INDUSTRIES SURGED AHEAD!

CINEMA
SCIENTIFICALLY AIR CONDITIONED

BUT WE'LL *ALSO* COME BACK TO THIS LATER...

IN THE EARLY 1940s, LOS ANGELES PROMOTED CARS OVER TROLLEYS...

THE LOCALS THOUGHT THE NEW POLLUTION CAME FROM JAPANESE GAS ATTACKS!

ARIE JAN HAAGEN-SMIT (DUTCH, 1900-77) PUBLISHED RESEARCH ON THIS *SMOG** BY 1952...

I CAN SMELL AND TASTE IT!

* SMOKE + FOG

PHOTOCHEMICAL SMOG WORKS LIKE THIS:

INSIDE CAR ENGINE, $N_2 + O_2 \longrightarrow 2NO$

CAR EXHAUST: $2NO + O_2 \longrightarrow 2NO_2$ (RED-BROWN GAS)

IN SUNLIGHT, $NO_2 + LIGHT \longrightarrow NO + O$

$O_2 + O \longrightarrow O_3$ (OZONE)

IRRITANT, DEGRADES RUBBER

ALSO GET HYDROCARBON FUEL + O $\longrightarrow$ R· + ·OH
(VAPOR) (RADICALS— CHAPTER 13)

EUGÈNE JULES HOUDRY (FRENCH-AMERICAN, 1892-1962) PATENTED A *CATALYTIC CONVERTER* IN 1956—IT REDUCED HYDROCARBON EXHAUST!

April 17, 1956
E. J. HOUDRY
CATALYTIC STRUCTURE AND TREATMENT
FILED SEPT. 28, 1953
2,742,437

BUT LEADED GASOLINE *POISONS* THE CATALYST!

BY THE 1930s, SCIENTISTS KNEW ABOUT THE ATMOSPHERE'S *OZONE LAYER*...

STRATOSPHERE
OZONE LAYER ————— 20–30 km ALTITUDE
TROPOSPHERE

IT PROTECTS LIFE BY ABSORBING UV FROM THE SUN
OZONE IS FORMED BY LIGHT:
O_2 + LIGHT $\longrightarrow$ 2O
O + O_2 $\rightleftharpoons$ O_3 (OZONE)

IN 1974, MARIO JOSÉ MOLINA-PASQUEL HENRÍQUEZ (MEXICAN, 1943–2020) & FRANK SHERWOOD ROWLAND (AMERICAN, 1927–2012) SHOWED HOW CFCs BREAK UP IN THE STRATOSPHERE...

CFCs BREAK APART *ALSO* VIA LIGHT:

$CFCl_3$ + LIGHT $\longrightarrow$ $\cdot CFCl_2$ + $\cdot Cl$
CCl_2F_2 + LIGHT $\longrightarrow$ $\cdot CF_2Cl$ + $\cdot Cl$

FORMING REACTIVE RADICALS!

HIGHLY REACTIVE CHLORINE ATOMS *THEN*...

$Cl\cdot$ + O_3 $\longrightarrow$ ClO + O_2
ClO + O $\longrightarrow$ $Cl\cdot$ + O_2

CHAIN REACTION (SEE CHAPTER 9) DESTROYING THE OZONE LAYER!

BY 1985, THREE BRITONS—JOSEPH CHARLES FARMAN (1930–2013), BRIAN GERARD GARDINER, AND JONATHAN SHANKLIN (b. 1953) FOUND THAT OZONE LEVELS WERE DROPPING FASTER OVER ANTARCTICA!

WE CALL IT THE *OZONE HOLE!*

0   100 200 300 400 500 600
Dobson units

THE BLUE AREA!

IN THREE AND A HALF YEARS, THE *MONTRÉAL PROTOCOL* BEGAN PHASING OUT INTERNATIONAL PRODUCTION OF CFCs...

MONTREAL
1989
PROTOCOL

SCIENTISTS ESTIMATE FULL RECOVERY OF THE OZONE LAYER IN ABOUT A CENTURY!

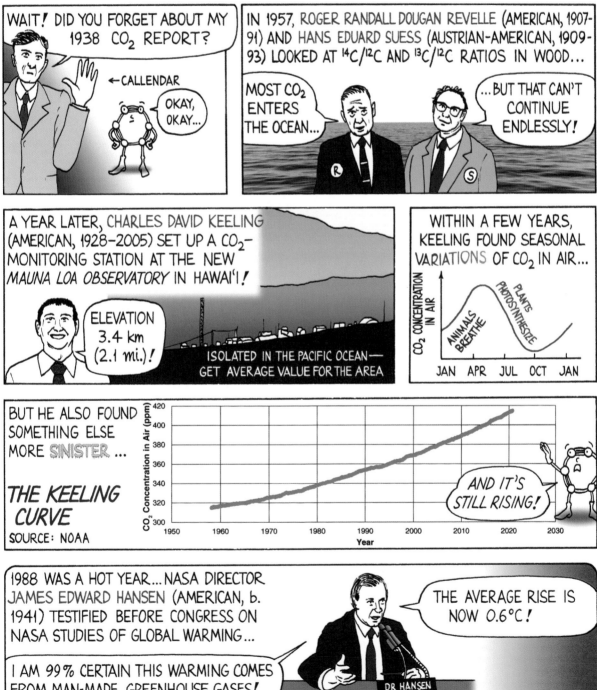

WAIT! DID YOU FORGET ABOUT MY 1938 CO$_2$ REPORT?

← CALLENDAR

OKAY, OKAY...

IN 1957, ROGER RANDALL DOUGAN REVELLE (AMERICAN, 1907-91) AND HANS EDUARD SUESS (AUSTRIAN-AMERICAN, 1909-93) LOOKED AT $^{14}C/^{12}C$ AND $^{13}C/^{12}C$ RATIOS IN WOOD...

MOST CO$_2$ ENTERS THE OCEAN...

...BUT THAT CAN'T CONTINUE ENDLESSLY!

A YEAR LATER, CHARLES DAVID KEELING (AMERICAN, 1928–2005) SET UP A CO$_2$-MONITORING STATION AT THE NEW *MAUNA LOA OBSERVATORY* IN HAWAI'I!

ELEVATION 3.4 km (2.1 mi.)!

ISOLATED IN THE PACIFIC OCEAN— GET AVERAGE VALUE FOR THE AREA

WITHIN A FEW YEARS, KEELING FOUND SEASONAL VARIATIONS OF CO$_2$ IN AIR...

CO$_2$ CONCENTRATION IN AIR

PLANTS PHOTOSYNTHESIZE

ANIMALS BREATHE

JAN APR JUL OCT JAN

BUT HE ALSO FOUND SOMETHING ELSE MORE SINISTER ...

THE KEELING CURVE

SOURCE: NOAA

CO$_2$ Concentration in Air (ppm)

420
400
380
360
340
320
300

1950 1960 1970 1980 1990 2000 2010 2020 2030

Year

AND IT'S STILL RISING!

1988 WAS A HOT YEAR... NASA DIRECTOR JAMES EDWARD HANSEN (AMERICAN, b. 1941) TESTIFIED BEFORE CONGRESS ON NASA STUDIES OF GLOBAL WARMING...

I AM 99% CERTAIN THIS WARMING COMES FROM MAN-MADE GREENHOUSE GASES!

THE AVERAGE RISE IS NOW 0.6°C!

DR HANSEN

STARTING IN 1990, WITH THE POLLUTION PREVENTION ACT, THE USA'S ENVIRONMENTAL PROTECTION AGENCY BEGAN RESEARCH INTO...

...REDESIGNING CHEMICAL PROCESSES TO LOWER ENVIRON-MENTAL EFFECTS...

"GREEN CHEMISTRY" FIRST APPEARED IN A PAPER IN 1990, BUT PAUL T. ANASTAS (AMERICAN, b. 1962) POPULARIZED IT IN THE 1990s ...

HE AND JOHN C. WARNER (AMERICAN, b. 1962) CAME UP WITH *12 PRINCIPLES OF GREEN CHEMISTRY* IN 1998, INCLUDING:

- *PREVENT WASTE INSTEAD OF CLEANING UP*
- *USE SAFER REAGENTS*
- *USE RENEWABLE RAW MATERIALS*
- *AVOID SOLVENTS*
- *FOR SYNTHESES, MINIMIZE WASTE — WHICH SHOULD BE THE LEAST TOXIC POSSIBLE*
- *BE ENERGY-EFFICIENT — AVOID HIGH PRESSURES AND TEMPERATURES*

AN EXAMPLE OF GREEN CHEMISTRY: MANUFACTURE OF POLYCARBONATE PLASTIC...

TYPICAL... $CO\ (g) + Cl_2\ (g) \longrightarrow$ CARBONYL CHLORIDE

POISONOUS CARBON MONOXIDE — CHLORINE GAS

THEN $n\,COCl_2$ DISSOLVED IN CH_2Cl_2 + n [...]$^{-2}$ DISSOLVED IN NaOH (aq) — POLYMERIZES WHERE SOLUTIONS MEET — POLYCARBONATE $+\ 2n\,Cl^-$

GREENER VERSION...

n HO— ... —OH + n O=C ... DIPHENYL CARBONATE

POLYCARBONATE $+\ 2n$ ... —OH

THE GREEN METHOD AVOIDS SO MANY TOXIC MOLECULES!

CHAPTER 17: NANOCHEMISTRY

A CHEMISTRYSCOPE PICTURE

21st CENTURY-FACTS

CHEMISTS GENERALLY MIX THINGS TOGETHER AND HOPE FOR THE BEST...

(WITH APPROPRIATE SAFETY EQUIPMENT)

BUT CAN WE WORK THE *OTHER* WAY... ATOM BY ATOM UPWARD?

I'M BAAAACK... DID SOMEONE SAY *"ATOM"*?

RICHARD PHILLIPS FEYNMAN (AMERICAN, 1918–88), CALIFORNIA INSTITUTE OF TECHNOLOGY, DECEMBER 1959

WHAT WOULD HAPPEN IF WE COULD ARRANGE THE ATOMS ONE BY ONE THE WAY WE WANT THEM?

HA HA! WHAT A *JOKESTER!*

HEY... DID YOU FORGET ABOUT MY *"MOLECULAR VALVES"* IN MY 1956 STORY, "THE LAST QUESTION"?

DR I. ASIMOV

IN 1974, NORIO TANIGUCHI (JAPAN, 1912–99) INVENTED THE WORD

NANO-TECHNOLOGY!

A *NANOMETER** (nm) MEANS ONE (AMERICAN) BILLIONTH OF A METER — ABOUT 8 OR 10 ATOMS LONG!

* GREEK νᾶνος, NANOS, "DWARF"

THE NANOWORLD INCLUDES SIZES FROM ABOUT 1 nm TO ABOUT 100 nm — THE SIZE OF MEDIUM TO HUGE MOLECULES!

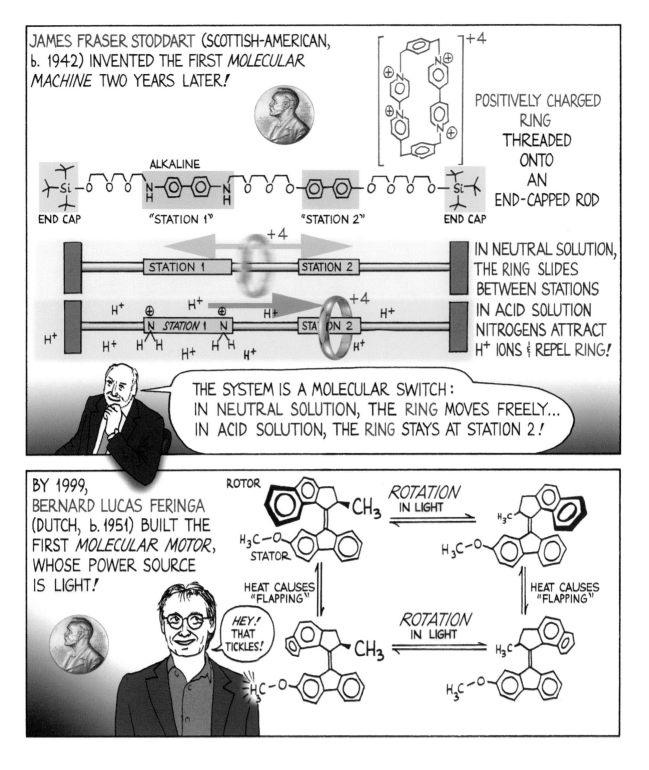

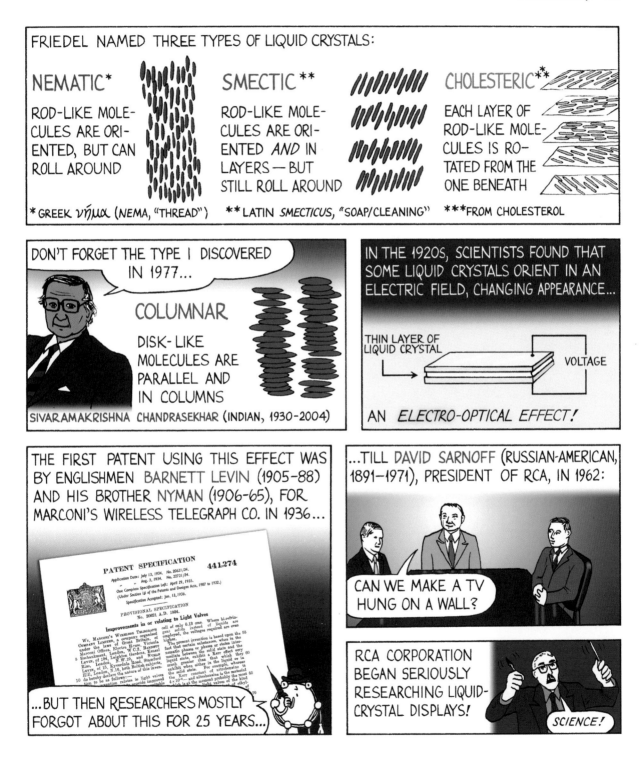

FRIEDEL NAMED THREE TYPES OF LIQUID CRYSTALS:

NEMATIC* — ROD-LIKE MOLECULES ARE ORIENTED, BUT CAN ROLL AROUND

SMECTIC** — ROD-LIKE MOLECULES ARE ORIENTED *AND* IN LAYERS — BUT STILL ROLL AROUND

CHOLESTERIC** — EACH LAYER OF ROD-LIKE MOLECULES IS ROTATED FROM THE ONE BENEATH

* GREEK νῆμα (NEMA, "THREAD") ** LATIN *SMECTICUS*, "SOAP/CLEANING" ***FROM CHOLESTEROL

DON'T FORGET THE TYPE I DISCOVERED IN 1977...

COLUMNAR — DISK-LIKE MOLECULES ARE PARALLEL AND IN COLUMNS

SIVARAMAKRISHNA CHANDRASEKHAR (INDIAN, 1930-2004)

IN THE 1920s, SCIENTISTS FOUND THAT SOME LIQUID CRYSTALS ORIENT IN AN ELECTRIC FIELD, CHANGING APPEARANCE...

THIN LAYER OF LIQUID CRYSTAL

VOLTAGE

AN *ELECTRO-OPTICAL EFFECT!*

THE FIRST PATENT USING THIS EFFECT WAS BY ENGLISHMEN BARNETT LEVIN (1905–88) AND HIS BROTHER NYMAN (1906-65), FOR MARCONI'S WIRELESS TELEGRAPH CO. IN 1936...

PATENT SPECIFICATION 441,274

...BUT THEN RESEARCHERS MOSTLY FORGOT ABOUT THIS FOR 25 YEARS...

...TILL DAVID SARNOFF (RUSSIAN-AMERICAN, 1891–1971), PRESIDENT OF RCA, IN 1962:

CAN WE MAKE A TV HUNG ON A WALL?

RCA CORPORATION BEGAN SERIOUSLY RESEARCHING LIQUID-CRYSTAL DISPLAYS!

SCIENCE!

INSTEAD OF A SOLID METAL ANODE, I USED AN ANODE MADE OF "PETROLEUM COKE," A TYPE OF LAYERED CARBON. SO NOW *BOTH* ELECTRODES ARE INTERCALATED!

HEXAGONAL ARRAYS OF C ATOMS IN LAYERS (COLORED SO YOU CAN SEE THE LAYERS MORE EASILY)

IT'S MORE STABLE AND IT'S LIGHTWEIGHT!

COMMERCIAL PRODUCTION OF THESE Li-ION BATTERIES BEGAN IN 1991...

PRETTY RAD, HUH?

BUT SPEAKING OF ELECTRO-CHEMISTRY, WHAT ABOUT DR. ASIMOV'S "MOLECULAR VALVES"?

FIRST LET'S EXAMINE "VALVE"...

ONE WAY

HOT FILAMENT

JOHN AMBROSE FLEMING (ENGLISH, 1849–1945) PATENTED THE FIRST ELECTRONIC VALVE—A *DIODE*—IN 1904!

LEE DEFOREST (AMERICAN, 1873–1961) IN 1906 ADDED A GRID, SO AMPLIFICATION OF A WEAK SIGNAL WAS POSSIBLE!

APPLY WEAK RADIO SIGNAL HERE

GRID

AMPLIFIED SIGNAL IS HERE

HOT FILAMENT

TRIODE

SIMULTANEOUSLY, THE SOLID-STATE SEMICONDUCTOR DIODE WAS USED TO DETECT RADIO WAVES...

Metal point
Silicon

August, 1919   ELECTRICAL EXPERIMENTER   325

How I Invented the Crystal Detector
By GREENLEAF WHITTIER PICKARD

Anode   Cathode

WITH BETTER SEMICONDUCTOR PROCESSING, THE *TRANSISTOR* (SOLID-STATE TRIODE) WAS INVENTED IN 1947!

E   AMPLIFIED SIGNAL IS HERE

B

C

APPLY WEAK RADIO SIGNAL HERE

JOHN BARDEEN
WILLIAM SHOCKLEY
WALTER BRATTAIN

EPILOGUE

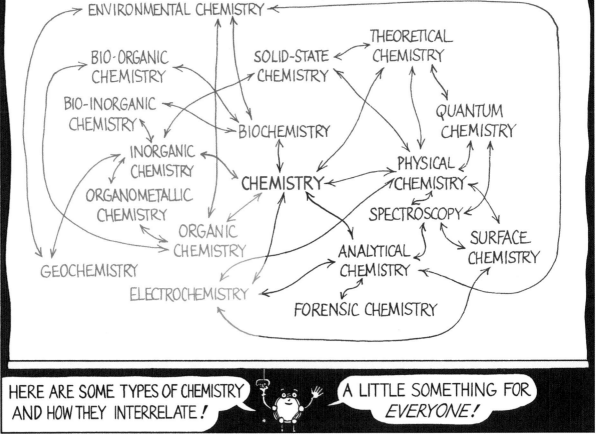

Index

Wien, Wilhelm, 138
Wilfarth, Hermann, 171
Wilkins, Maurice, 158
Williamson, Alexander William, 65, 93, 94
Williamson, Catherine, 94
Winkler, Clemens, 75
Wöhler, Friedrich, 61, 63, 64, 66
Wood, 6, 20, 31, 33, 34, 37, 64, 111, 177
wood naphtha. *See* methanol.
World War I, 122, 123, 124, 137
World War II, 126, 157
Woulfe, Peter, 111
Wright, Joseph, 25
Wurtz, Adolphe, 66, 79, 84

xanthosis, 12
xenon, 78, 182
    salt, 147

x ray, 131, 133, 136, 139
    characteristic, 136, 137
    crystallography, 136, 154, 158
Xylonite, 151

yeast, 100
Yiddish, 123
Yin-Yang, 6
Yoshino, Akira, 188
Youmans, Edward L., 67
yours truly, 182, 190

Zeitschrift für Physikalische Chemie, 105
Zeno of Citium, 5
Ziegler, Karl, 158
zinc, 24, 55, 56, 68
    sulfide, 136
zirconium salt, 158

Dmitri Mendeleev

Stefanie Horovitz

Gilbert N. Lewis

Mario J. Molina

Sivaramakrishna Chandrasekhar

Marie Curie

Rosalind Franklin

John Dalton

Ben Zene, Your host & guide

Antoine Lavoisier

Democritos